Selected Titles in This Series

717 **Ron Brown,** Frobenius groups and classical maximal orders, 2001

716 **John H. Palmieri,** Stable homotopy over the Steenrod algebra, 2001

715 **W. N. Everitt and L. Markus,** Multi-interval linear ordinary boundary value problems and complex symplectic algebra, 2001

714 **Earl Berkson, Jean Bourgain, and Aleksander Pełczynski,** Canonical Sobolev projections of weak type $(1,1)$, 2001

713 **Dorina Mitrea, Marius Mitrea, and Michael Taylor,** Layer potentials, the Hodge Laplacian, and global boundary problems in nonsmooth Riemannian manifolds, 2001

712 **Raúl E. Curto and Woo Young Lee,** Joint hyponormality of Toeplitz pairs, 2001

711 **V. G. Kac, C. Martinez, and E. Zelmanov,** Graded simple Jordan superalgebras of growth one, 2001

710 **Brian Marcus and Selim Tuncel,** Resolving Markov chains onto Bernoulli shifts via positive polynomials, 2001

709 **B. V. Rajarama Bhat,** Cocylces of CCR flows, 2001

708 **William M. Kantor and Ákos Seress,** Black box classical groups, 2001

707 **Henning Krause,** The spectrum of a module category, 2001

706 **Jonathan Brundan, Richard Dipper, and Alexander Kleshchev,** Quantum Linear groups and representations of $GL_n(\mathbb{F}_q)$, 2001

705 **I. Moerdijk and J. J. C. Vermeulen,** Proper maps of toposes, 2000

704 **Jeff Hooper, Victor Snaith, and Min van Tran,** The second Chinburg conjecture for quaternion fields, 2000

703 **Erik Guentner, Nigel Higson, and Jody Trout,** Equivariant E-theory for C^*-algebras, 2000

702 **Ilijas Farah,** Analytic guotients: Theory of liftings for quotients over analytic ideals on the integers, 2000

701 **Paul Selick and Jie Wu,** On natural coalgebra decompositions of tensor algebras and loop suspensions, 2000

700 **Vicente Cortés,** A new construction of homogeneous quaternionic manifolds and related geometric structures, 2000

699 **Alexander Fel'shtyn,** Dynamical zeta functions, Nielsen theory and Reidemeister torsion, 2000

698 **Andrew R. Kustin,** Complexes associated to two vectors and a rectangular matrix, 2000

697 **Deguang Han and David R. Larson,** Frames, bases and group representations, 2000

696 **Donald J. Estep, Mats G. Larson, and Roy D. Williams,** Estimating the error of numerical solutions of systems of reaction-diffusion equations, 2000

695 **Vitaly Bergelson and Randall McCutcheon,** An ergodic IP polynomial Szemerédi theorem, 2000

694 **Alberto Bressan, Graziano Crasta, and Benedetto Piccoli,** Well-posedness of the Cauchy problem for $n \times n$ systems of conservation laws, 2000

693 **Doug Pickrell,** Invariant measures for unitary groups associated to Kac-Moody Lie algebras, 2000

692 **Mara D. Neusel,** Inverse invariant theory and Steenrod operations, 2000

691 **Bruce Hughes and Stratos Prassidis,** Control and relaxation over the circle, 2000

690 **Robert Rumely, Chi Fong Lau, and Robert Varley,** Existence of the sectional capacity, 2000

689 **M. A. Dickmann and F. Miraglia,** Special groups: Boolean-theoretic methods in the theory of quadratic forms, 2000

688 **Piotr Hajłasz and Pekka Koskela,** Sobolev met Poincaré, 2000

(Continued in the back of this publication)

Frobenius Groups and Classical Maximal Orders

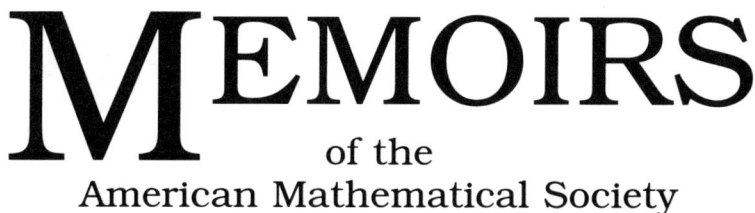
of the
American Mathematical Society

Number 717

Frobenius Groups and Classical Maximal Orders

Ron Brown

May 2001 • Volume 151 • Number 717 (third of 5 numbers) • ISSN 0065-9266

American Mathematical Society
Providence, Rhode Island

2000 *Mathematics Subject Classification.*
Primary 20E99; Secondary 20B20, 11R04, 11R20, 16H05, 16K20, 20F05.

Library of Congress Cataloging-in-Publication Data
Brown, Ron, 1943–
  Frobenius groups and classical maximal orders / Ron Brown.
    p. cm. — (Memoirs of the American Mathematical Society, ISSN 0065-9266 ; no. 717)
  "Volume 151, number 717 (third of 5 numbers)."
  Includes bibliographical references.
  ISBN 0-8218-2667-0 (alk. paper)
  1. Frobenius groups. I. Title. II. Series.
QA3.A57 no. 717
[QA177]
510s—dc21]
[512′.2]   2001018230

Memoirs of the American Mathematical Society

This journal is devoted entirely to research in pure and applied mathematics.

Subscription information. The 2001 subscription begins with volume 149 and consists of six mailings, each containing one or more numbers. Subscription prices for 2001 are $494 list, $395 institutional member. A late charge of 10% of the subscription price will be imposed on orders received from nonmembers after January 1 of the subscription year. Subscribers outside the United States and India must pay a postage surcharge of $31; subscribers in India must pay a postage surcharge of $43. Expedited delivery to destinations in North America $35; elsewhere $130. Each number may be ordered separately; *please specify number* when ordering an individual number. For prices and titles of recently released numbers, see the New Publications sections of the *Notices of the American Mathematical Society*.

Back number information. For back issues see the *AMS Catalog of Publications*.

Subscriptions and orders should be addressed to the American Mathematical Society, P. O. Box 845904, Boston, MA 02284-5904. *All orders must be accompanied by payment.* Other correspondence should be addressed to Box 6248, Providence, RI 02940-6248.

Copying and reprinting. Individual readers of this publication, and nonprofit libraries acting for them, are permitted to make fair use of the material, such as to copy a chapter for use in teaching or research. Permission is granted to quote brief passages from this publication in reviews, provided the customary acknowledgment of the source is given.

Republication, systematic copying, or multiple reproduction of any material in this publication is permitted only under license from the American Mathematical Society. Requests for such permission should be addressed to the Assistant to the Publisher, American Mathematical Society, P. O. Box 6248, Providence, Rhode Island 02940-6248. Requests can also be made by e-mail to reprint-permission@ams.org.

Memoirs of the American Mathematical Society is published bimonthly (each volume consisting usually of more than one number) by the American Mathematical Society at 201 Charles Street, Providence, RI 02904-2294. Periodicals postage paid at Providence, RI. Postmaster: Send address changes to Memoirs, American Mathematical Society, P. O. Box 6248, Providence, RI 02940-6248.

© 2001 by the American Mathematical Society. All rights reserved.
This publication is indexed in *Science Citation Index*®, *SciSearch*®, *Research Alert*®, *CompuMath Citation Index*®, *Current Contents*®/*Physical, Chemical & Earth Sciences*.
Printed in the United States of America.

∞ The paper used in this book is acid-free and falls within the guidelines established to ensure permanence and durability.
Visit the AMS home page at URL: http://www.ams.org/

10 9 8 7 6 5 4 3 2 1   06 05 04 03 02 01

Contents

Chapter 1.	Introduction	1
Chapter 2.	Lemmas on Truncated Group Rings	6
Chapter 3.	Groups of Real Quaternions	10
Chapter 4.	Proof of the Classification Theorem	14
Chapter 5.	Frobenius complements with core index 1	18
Chapter 6.	Frobenius complements with core index 4	21
Chapter 7.	Frobenius complements with core index 12	31
Chapter 8.	Frobenius complements with core index 24	42
Chapter 9.	Frobenius complements with core index 60	48
Chapter 10.	Frobenius complements with core index 120	50
Chapter 11.	Counting Frobenius Complements	54
Chapter 12.	Maximal Orders	57
Chapter 13.	Isomorphism Classes of Frobenius Groups with Abelian Frobenius Kernel	66
Chapter 14.	Concrete Constructions of Frobenius Groups	76
Chapter 15.	Counting Frobenius Groups with Abelian Frobenius Kernel	84
Chapter 16.	Isomorphism Invariants for Frobenius Complements	96
Chapter 17.	Schur Indices and Finite Subgroups of Division Rings	102
Bibliography		110

Abstract

The analysis of the set of isomorphism classes of Frobenius groups with commutative Frobenius kernel is reduced here to "abelian" algebraic number theory. Some problems, such as the computation of the number of isomorphism classes of Frobenius groups subject to various restrictions on orders, are further reduced to elementary number theory. The starting point is the bijection between the set of isomorphism classes of Frobenius groups with commutative Frobenius kernel and with given Frobenius complement G and the set of G–semi-linear isomorphism classes of finite modules over a ring naturally associated with G. This ring is a maximal order in a simple algebra whose center $\mathcal{Z}$ is an abelian extension of $\mathbb{Q}$. All Frobenius complements and their associated rings are explicitly computed here in terms of simple numerical invariants. The finite modules of such a ring are sums of indecomposable ones, and the indecomposable ones are shown to correspond to powers of unramified (over $\mathbb{Q}$) maximal ideals of the ring of integers of $\mathcal{Z}$ which do not contain the order of the Frobenius complement.

Received by the editor August 3, 1998.
2000 *Mathematics Subject Classification.* Primary 20E99,
Secondary 20B20, 11R04, 11R20, 16H05, 16K20, 20F05.
Key words and phrases. Frobenius group, Frobenius complement, Frobenius kernel, truncated group ring, maximal order, central simple algebra, abelian number field, Schur index.

CHAPTER 1

Introduction

A *Frobenius group* is a finite group G with a nontrivial normal subgroup N (called a *Frobenius kernel*) and a nontrivial subgroup H (called a *Frobenius complement*) such that the orders of N and of H are relatively prime and for every $x \in G \backslash N$ there exists a unique $y \in N$ with $x \in yHy^{-1}$. Particular classes of Frobenius groups are of interest (see [**J1**, Example, pp. 320–324] for a striking application), as is the class of Frobenius groups as a whole (e.g., proofs of general results involving minimal counterexamples sometimes reduce to an analysis of Frobenius groups and their automorphism groups). The structure of Frobenius groups has been clarified by deep results such as Thompson's theorem (settling a long–standing conjecture) that Frobenius kernels are nilpotent [**T**, Theorem 1, p. 579] and Zassenhaus's structure theory for Frobenius complements [**Pa**], [**Z**].

The study of Frobenius groups with abelian Frobenius kernel is reduced in this paper to algebraic number theory. The author and D. K. Harrison [**BrH**] showed that there is associated with each Frobenius complement H a ring, called its truncated group ring (cf. Chapter 2), such that the G–semi-linear isomorphism classes (cf. Remark 13.3) of finite modules over that ring are naturally bijective with the isomorphism classes of Frobenius groups with Frobenius complement H and abelian Frobenius kernel. (Frobenius kernels are *always* abelian except when the complement is a group of odd order all of whose Sylow subgroups are cyclic. Of the 5,385,907 isomorphism classes of Frobenius complements of order at most 10^6 only about 11% are of this type. Combinatorial results for Frobenius complements are discussed in Chapter 11.) For example, if G is a metabelian Frobenius group, then its Frobenius complement is cyclic, say of order m, and the truncated group ring of the complement is $\mathbb{Z}[1/m, e^{2\pi i/m}]$. Since finite modules over such rings are well–understood, one can obtain a very complete picture of the class of metabelian Frobenius groups, including counting theorems for isomorphism classes (cf. [**BrH**, Section 11]). Another example of a Frobenius complement is the multiplicative group of real quaternions $\{\pm 1, \pm \mathbf{i}, \pm \mathbf{j}, \pm \mathbf{k}\}$ (the *quaternion group*). Here the truncated group ring is the localization $\mathbb{Z}[\mathbf{i}, \mathbf{j}, \mathbf{k}, \frac{1}{2}]$ of the Hurwitz ring [**H**, p. 373].

In this paper a complete computation of the truncated group rings of arbitrary Frobenius complements is given in terms of simple numerical invariants which determine the isomorphism classes of the Frobenius complements. As the next theorem indicates, these rings are of such a nature that the problem of computing their finite modules can be regarded as a problem in algebraic number theory, albeit in a broad sense.

1.1. THEOREM. *The truncated group ring of a Frobenius complement of order γ is a maximal $\mathbb{Z}[1/\gamma]$–order in a finite dimensional simple algebra whose center is an abelian Galois field extension of $\mathbb{Q}$.*

The above theorem lets us apply the theory of modules over maximal orders to find the isomorphism classes of Frobenius groups with abelian Frobenius kernel.

1.2. THEOREM. *There is a natural bijection from the set of isomorphism classes of Frobenius groups with abelian Frobenius kernel and with given Frobenius complement G to the set of orbits (under the natural action of $\operatorname{Aut} G$) of the free abelian semigroup on the set of all powers of maximal ideals of the center of the truncated group ring of G.*

In Chapter 13 we prove a more complete statement of Theorem 1.2, namely Theorem 13.2. We also give there a simple description of all abelian Frobenius kernels and find the thirteen isomorphism classes of nonsolvable Frobenius groups of order at most one million. The Frobenius groups of order at most one billion with noncyclic Frobenius complement and *nonabelian* Frobenius kernel are shown to have only three possible orders. The results of Chapter 13 are applied in the next one to give concrete constructions of all Frobenius groups with abelian Frobenius kernel. These are shown in Theorem 14.2 to be semidirect products (see below)

$$\left(\bigoplus_{\mathfrak{a}} I'/\mathfrak{a}I'\right) \rtimes H$$

where H is a Frobenius complement, the $\mathfrak{a}$ are powers of maximal ideals in a ring I, and I and I' are the integral closures, respectively, of $\mathbb{Z}[1/|H|]$ in the center of and in a canonical maximal subfield of the simple algebra of Theorem 1.1 arising from H. (If a group B (written multiplicatively) acts on a group A (written additively), then we let $A \rtimes B$ denote the *semidirect product*, i.e., the set $A \times B$ with the operation given by the formula $(a, b)(c, d) = (a + bc, bd)$.) Constructions of all Frobenius groups whose Frobenius complement is the special linear group $SL(2, 5)$ or a binary dihedral group are given involving little more than elementary number theory. Here is a simple example.

1.3. EXAMPLE. In Remark 14.5 below we show that the Frobenius groups whose Frobenius complement is the quaternion group $\langle \mathbf{i}, \mathbf{j} \rangle$ are up to isomorphism exactly the semidirect products $H := M \rtimes \langle \mathbf{i}, \mathbf{j} \rangle$ where M is a direct sum (uniquely determined by the isomorphism class of H except for the order of the factors) of groups of the form $M(p^k) := (\mathbb{Z}/p^k\mathbb{Z})^2$ where p is an odd prime, $k > 0$, and the action of $\langle \mathbf{i}, \mathbf{j} \rangle$ on $M(p^k)$ is such that for all $(\delta, \gamma) \in M(p^k)$ we have $\mathbf{i}(\delta, \gamma) = (a\delta + b\gamma, b\delta - a\gamma)$ and $\mathbf{j}(\delta, \gamma) = (\gamma, -\delta)$ where for each p^k we have fixed integers a and b with $a^2 + b^2 \equiv -1 \pmod{p^k}$.

Theorem 1.2 is applied in Chapter 15 to give a formula (Theorem 15.1) in terms of elementary number theory for the number of isomorphism classes of Frobenius groups with abelian Frobenius kernel of given order and with given Frobenius complement. The formula is applied to calculate, for example, that there are exactly 569,342 isomorphism classes of such groups of order at most one million.

Theorem 1.1 will be proved in Chapter 12. The proof will depend on the analysis in Chapters 5 through 10 of six types of Frobenius complements. The breakup into six cases arises from the following sharpening of a theorem of Zassenhaus [**Pa**, Theorem 18.2, p. 196] on the structure of solvable Frobenius complements.

1.4. CLASSIFICATION THEOREM. *Every Frobenius complement G has a unique normal subgroup N such that all Sylow subgroups of N are cyclic and G/N is* *isomorphic to one of the following six groups:*

(1) $\qquad\qquad\qquad 1,\quad V_4,\quad A_4,\quad S_4,\quad A_5,\quad S_5$

where 1 *denotes the trivial group and* V_4 *denotes the Sylow* 2*–subgroup of the alternating group* A_4.

Theorem 1.4 will be proved in Chapter 4 using the results of the preceding two chapters. In Chapter 2 we will review the construction of the truncated group ring of a group and prove some basic lemmas about them, including an analysis of the truncated group rings of direct products and of cyclic extensions. (This chapter has references to [**BrH**], but it, and in fact almost all of this paper, can be read independently of [**BrH**]; we do include for completeness a few short arguments from that paper.) The truncated group rings of finite subgroups of the multiplicative group of units of the division ring of real quaternions are computed in Chapter 3. These groups give standard examples of five of the six types of Frobenius complements and they play a special role in the general theory. We also lay the foundation in Chapter 3 for the construction in Chapter 4 of an example of the sixth type of Frobenius complement.

We referred earlier to numerical invariants determining the isomorphism class of a Frobenius complement. Detailed insights into the structure of Frobenius complements and related groups can be found in the literature. The work here inevitably overlaps some of this (e.g., Theorems 7 and 16 of [**Z**] give presentations by generators and relations related to ones given here), but it is distinctive in the identification of numerical invariants which determine isomorphism classes with sufficient precision to allow one to read off the number of isomorphism classes of Frobenius complements of various sorts. (The counting theorems here are less interesting to me in themselves than as tests of the completeness of the theory.)

The next definition introduces one group of invariants which determines a Frobenius complement up to isomorphism. The definition uses the fact (easily deduced from [**S**, 12.6.17, p. 356]) that if N is a group all of whose Sylow subgroups are cyclic (*i.e.*, N is a $\mathbb{Z}$*-group* [**Pa**, p. 104]), then the group $N' \times (N/N')$ is cyclic, so its automorphism group is canonically isomorphic to $\mathbb{Z}_{|N|}^{\bullet}$. ($N'$ denotes the commutator subgroup of N.)

1.5. DEFINITION. Let G and N be as in Theorem 1.4. We call N the *core* of G and $[G : N]$ the *core index* of G. The *signature* of G is the image of the composition of maps

$$G \longrightarrow \operatorname{Aut}\left(N' \times (N/N')\right) \longrightarrow \mathbb{Z}_{|N|}^{\bullet}$$

where the left-hand map is induced by conjugation by elements of G and the right-hand map is the canonical isomorphism.

1.6. THEOREM. *A Frobenius complement is determined up to isomorphism by its order, core index, and signature.*

Theorem 1.6 will be proved in Chapter 16, where we will also show that other combinations of invariants—especially ones arising from the truncated group ring—determine the isomorphism class of a Frobenius complement. For example, in the statement of the above theorem the signatures can be replaced by the abelian extensions (of the rational numbers) from Theorem 1.1. The arguments in Chapters 11 through 17 will depend on the analysis of each of the six types of Frobenius complements, namely those with core index 1, 4, 12, 24, 60 and 120 (the orders of the six groups in the list (1)). In Chapters 5 through 10 we will determine one after another the set of isomorphism classes and the truncated group rings of Frobenius complements with given core index in terms of numerical invariants related to the signatures above but more specifically tailored to the specific type of Frobenius complement. In each case we will start with some numerical data, use it to construct a ring and to identify a subgroup of the group of units of the ring, and then prove that the group is a Frobenius complement, the ring is its truncated group ring, the numerical data are isomorphism invariants of the group, and all Frobenius complements with the given core index are constructed in this way. This approach is focused specifically on Frobenius complements since these are precisely the nontrivial groups whose truncated group rings are nontrivial [**BrH**, Theorem 8.4].

The method of constructing groups sketched above is related to Amitsur's construction [**Am**] of the groups which are isomorphic to finite subgroups of the groups of units of division rings. This should not be surprising since all such groups are known to be Frobenius complements [**SW**, Theorem 2.1.2, p. 45]. The precise overlap of this paper with [**Am**] is obscured by the focus here on a systematic (indeed, functorial) method of associating rings to groups by means of truncated group rings. In both cases the association allows the application of arithmetic methods. Such methods are used in Theorem 17.4 to give a formula involving only elementary number theory for the index of the central simple algebra of a Frobenius complement which is a $\mathbb{Z}$-group. This formula is used to show that every Frobenius complement which is a $\mathbb{Z}$-group of order at most 100,000 can be expressed as group of $k \times k$-matrices over a division ring where the average value of k for these groups is less than 2. The formula is presented as a contribution to what would be a natural extension of Amitsur's work [**Am**], namely, the computation of the index and the degree of the central simple algebra associated with each Frobenius complement. (The degrees and centers of these algebras are computed in Chapters 5 through 10; a comparison of the degree and index shows how close a Frobenius complement is to being a subgroup of a division ring.) The formula is also used to establish a variant of Shirvani's characterization of the $\mathbb{Z}$-groups which are finite subgroups of division rings.

We end this chapter by collecting for reference some notational conventions; all those which are not standard will be explained again when first used.

For any integer n we set $\zeta_n = e^{2\pi i/n}$ and let n_0 denote the product of the distinct rational primes dividing n. $\mathbb{Z}$, $\mathbb{Q}$, $\mathbb{Z}_n$, and $\mathbb{Z}_n^{\bullet}$ denote the ring of integers, the ring of rationals, the factor ring $\mathbb{Z}/n\mathbb{Z}$, and its group of multiplicative units, respectively. The Euler phi-function is denoted by ϕ.

For any finite set S, unitary ring R, finite group G and $g \in G$ we let $|S|$ denote the number of elements in S, $|g|$ denote the order of g, G' denote the commutator subgroup of G, $\mathcal{Z}(G)$ denote the center of G, $R^{\bullet}$ denote the multiplicative group of units of R, $M_n(R)$ denote the ring of $n \times n$ matrices over R, and RG denote the group ring of G over R. We let $\langle g \rangle$ denote the subgroup of G generated by g, and use similar notation for the subgroup generated by a sequence of elements of G. The restriction of a function f with domain S to a subset T of S is denoted $f|T$. We let $\mathrm{Int}_F R$ denote the integral closure in a field F of a subring R. If σ is a function whose domain contains S, then $S^{\sigma} = \{a \in S : \sigma(a) = a\}$; similar notation is used for the fixed subset of S of a set of functions with domain containing S. Finally, we write $S \triangleleft G$ if S is a normal subgroup of G.

1. INTRODUCTION

Another group of conventions will prove convenient. Suppose g is an element of order n of a multiplicative group. It is common and unambiguous to write $g^{s+n\mathbb{Z}}$ for g^s (where $s \in \mathbb{Z}$). We will extend this notation to cosets whose moduli are multiples of n and even mix different moduli, e.g.,

$$g^{(3+n\mathbb{Z})+(5+2n\mathbb{Z})} = g^8.$$

We will similarly mix integers and cosets of various sorts in congruences: if a and b are either integers or cosets whose moduli are multiples of n, then we shall write $a \equiv b \pmod{n}$ if and only if $a + n\mathbb{Z} = b + n\mathbb{Z}$. For example, $1 + 12\mathbb{Z} \equiv 13 \equiv 7 + 6\mathbb{Z} \pmod 6$.

Next, if n is a multiple of an integer m, then for any integer a we unambiguously write $(a + n Z, m)$ for the greatest common divisor (a, m). Thus for example $2 = (6, 8) = (6 + 24\mathbb{Z}, 8)$.

Two other conventions will be used constantly. If m and n are positive integers, then $m//n$ will denote the largest factor of m relatively prime to n, and n_m will denote the largest factor of m dividing a power of n. Note $m = (n_m)(m//n)$. For example, $6_{120} = 24$ and $120//6 = 5$. We make the order of operation conventions so that $ab//cd = (ab)//(cd)$ and $ab_n = a(b_n)$.

When defining a symbol a to be an expression E, we often write $a := E$.

APPENDIX. Here is a list indicating the page on which is first introduced a notation (other than those given immediately above) or a definition which might be used in a *subsequent* chapter with little or no reminder.

page 1: Frobenius group, Frobenius kernel, Frobenius complement

page 2: $A \rtimes B$ (semidirect product)

page 3: V_4, $\mathbb{Z}$-group, core, core index, signature

page 6: $\widehat{g}$, $\mathfrak{a}_G$, $\overline{g}$, $\mathbb{Z}_{(G)}$, the integral truncated group ring $\mathbb{Z}\langle G \rangle$, the rational truncated group ring $\mathbb{Q}\langle G \rangle$, the truncated group ring $\mathbb{Z}_{(G)}\langle G \rangle$

page 8: (A, σ, m, c), $\widehat{\sigma}$, $\mathbb{Q}A$

page 10: $\mathbb{H}$, $\mathbb{R}$, $\boldsymbol{\alpha}$, $\boldsymbol{\beta}$, C_n, D_{4n}, H_{24}, H_{48}, H_{120}, γ^* (conjugate of γ)

page 14: generalized quaternion group

page 17: H_{240}

page 18: (proper) Frobenius triple, 1–complement; r–sequence and invariant of a 1–complement

page 21: core invariant, type, J-complement (for $J = V_4, A_4, S_4, A_5, S_5$); r–sequence, invariant and reduced invariant for a V_4–complement

page 31: r–sequence, invariant and reduced invariant for an A_4–complement

page 42:. r–sequence, invariant and reduced invariant for an S_4–complement

page 50: invariant and reduced invariant for an S_5–complement

page 62: $(E/K, \Phi)$ (crossed product algebra)

page 66: G–group, $Iso(G)$, $\mathcal{S}(G)$, $[\eta]$ (for $\eta \in \mathcal{S}(G)$)

CHAPTER 2

Lemmas on Truncated Group Rings

Let G be a finite group.

2.1. NOTATION. Let A be a subring of the field of rational numbers $\mathbb{Q}$. If $g \in G$ let $\hat{g}$ denote the image of g under the natural map from G into the group ring AG. Let $\mathfrak{a}_G$ denote the set of elements of AG of the form $\sum_{h \in \langle g \rangle} \hat{h}$ where $g \in G$ and $g \neq 1$. Also let $A\langle G \rangle = AG/(\mathfrak{a}_G)$ and $A_{(G)} = A[1/|G|]$. Finally for each $g \in G$, let $\overline{g} = \hat{g} + (\mathfrak{a}_G)$ be its image in $A\langle G \rangle$. (Note that the precise meanings of $\mathfrak{a}_G$ and of the ideal $(\mathfrak{a}_G)$ depend on the choice of A.)

In this paper we use the above notation only with A equal to either $\mathbb{Z}$, $\mathbb{Z}_{(G)}$ or $\mathbb{Q}$. We call $\mathbb{Z}\langle G \rangle$ and $\mathbb{Q}\langle G \rangle$ the *integral truncated group ring* and the *rational truncated group ring* of G, respectively. The ring $\mathbb{Z}_{(G)}\langle G \rangle$ is called *the truncated group ring* of G. (By [**BrH**, Lemma 8.1, p. 64] this ring is the same as the "truncated group ring" of [**BrH**, Definition 7.3, p. 63].)

In this chapter we focus on integral truncated group rings because of their apparent simplicity and the fact that the other truncated group rings are obtained from them essentially by tensoring with either $\mathbb{Z}_{(G)}$ or $\mathbb{Q}$.

Guralnick and Wiegand prove that the ideal $(\mathfrak{a}_G)$ of the integral group ring $\mathbb{Z}G$ has trivial intersection with $\mathbb{Z}$ if and only if G is a Frobenius complement [**GW**, Theorem 2.2, p. 570]. We will use several times the following corollary of this result; for completeness a proof is given using some arguments from [**BrH**].

2.2. LEMMA. *Suppose G is nontrivial. Then G is a Frobenius complement if $\mathbb{Z}\langle G \rangle$ is not a torsion $\mathbb{Z}$-module.*

PROOF. Let p be any rational prime not dividing $|G|$. The canonical image M of $\mathbb{Z}\langle G \rangle$ in $\mathbb{Q}\langle G \rangle$ is finitely generated and hence free as a $\mathbb{Z}$-module, and by hypothesis it is nontrivial. Hence M/pM is finite and nontrivial. Now suppose g is a nontrivial element of G; then g has a power g^s of prime order, say $q = |g^s|$. The canonical map $\mathbb{Z}[x] \longrightarrow \mathbb{Z}\langle G \rangle$ taking x to $\overline{g}^s$ takes the cyclotomic polynomial $\Phi_q(x) = 1 + x + x^2 + \cdots + x^{q-1}$ to zero (cf. Notation 2.1), and hence it induces a homomorphism $\mathbb{Z}[\zeta_q] \longrightarrow \mathbb{Z}\langle G \rangle$ taking ζ_q to $\overline{g}^s$. Since $q = \Phi_q(1) = \prod_{i=1}^{q-1}(1 - \zeta_q^i)$, therefore $\prod_{i=1}^{q-1}(1 - \overline{g}^{si}) = q$. Thus if $\overline{g}m = m$ for some $m \in M/pM$, then $\overline{g}^s m = m$ and so

$$0 = (1 - \overline{g}^s)m = \prod_{i=1}^{q-1}(1 - \overline{g}^{si})m = qm.$$

Since we also have $pm = 0$ and $p \neq q$, then $m = 0$. This shows that the natural action of G on M/pM is without fixed points. It follows easily using the definition of a Frobenius group in [**S**, p. 348] that the semidirect product $M/pM \rtimes G$ with respect to the natural action of G on M/pM is a Frobenius group with Frobenius complement isomorphic to G. □

We will see by inspection from the explicit computation in Chapters 5 through 10 of the integral truncated group rings of all Frobenius complements that the integral truncated group rings of Frobenius complements are actually nontrivial and free as $\mathbb{Z}$-modules (cf. Theorem 12.2). Until this is proved we will often as a matter of bookkeeping be either hypothesizing or noting that various integral truncated group rings are nontrivial and free as $\mathbb{Z}$-modules.

It will sometimes be more convenient to work with the image of G in $\mathbb{Z}\langle G \rangle$ rather than with G itself. The following lemma is then relevant.

2.3. LEMMA. *If $\mathbb{Z}\langle G \rangle$ is nontrivial and free as a $\mathbb{Z}$-module, then the natural map $G \longrightarrow \mathbb{Z}\langle G \rangle$ is injective.*

PROOF. Suppose $h \in G$ has order $s > 1$ and $\overline{h} = 1$. Then $s1 = \sum_{i=0}^{s-1} \overline{h}^i = 0$, since $\sum_{g \in \langle h \rangle} \widehat{g} \in \mathfrak{a}_G$. This contradicts the hypothesis that $\mathbb{Z}\langle G \rangle$ is nontrivial and free. □

The next lemma will be used in Chapter 3 in the analysis of groups of real quaternions and in Chapter 17 in the analysis of finite subgroups of division rings.

2.4. LEMMA. *Let G be a finite subgroup of the group of units $D^\bullet$ of a division ring D and let $\varphi : \mathbb{Z}G \longrightarrow D$ be the map induced by the inclusion map $G \longrightarrow D$. Then $\mathfrak{a}_G$ is contained in $\ker \varphi$.*

PROOF. Suppose $g \in G$ has order $n > 1$. Then $g - 1$ is a unit in D. Since $0 = g^n - 1 = (g-1)(1 + g + \cdots + g^{n-1})$, then $\sum_{h \in \langle g \rangle} \widehat{h} \in \ker \varphi$. Thus $\mathfrak{a}_G \subset \ker \varphi$. □

We describe below how to compute the truncated group rings of direct products and cyclic extensions of groups in terms of the truncated group rings of these groups. In the former case, the key construction is tensor product; in the latter it is an old construction of Albert. Albert states his result for algebras over fields, but it generalizes transparently to algebras over arbitrary commutative rings.

2.5. THEOREM. [**A**, Theorems 10 and 11, pp. 183–184]. *Let A be an algebra over a unitary commutative ring R. Let m be a positive integer, σ be an automorphism of A, and c be a unit of A with $\sigma(c) = c$ and $ca = \sigma^m(a)c$ for all $a \in A$. Then A is a subalgebra of an R-algebra B which has an element b such that $1, b, \ldots, b^{m-1}$ is a basis for B as a left A-module, $b^m = c$, and $ba = \sigma(a)b$ for all $a \in A$. Moreover, if A is simple with center K and $\sigma|K$ has order m, then B is simple with center K^σ.*

In the above theorem, $\sigma|K$ denotes the restriction of σ to K and K^σ is the fixed field of $\sigma|K$.

2.6. REMARK AND NOTATION. We describe here some identifications that will be used throughout the paper in order to keep notational complexity under control.

(A) Let G be a group. We often treat the natural isomorphisms $\mathbb{Z}_{(G)}\langle G \rangle \longrightarrow \mathbb{Z}_{(G)} \otimes \mathbb{Z}\langle G \rangle$ and $\mathbb{Q}\langle G \rangle \longrightarrow \mathbb{Q} \otimes \mathbb{Z}_{(G)}\langle G \rangle$ as identifications. Now suppose $\mathbb{Z}\langle G \rangle$ is free as a $\mathbb{Z}$-module. Then the canonical homomorphisms $\mathbb{Z}\langle G \rangle \longrightarrow \mathbb{Z}_{(G)} \otimes \mathbb{Z}\langle G \rangle$ and $\mathbb{Z}_{(G)}\langle G \rangle \longrightarrow \mathbb{Q} \otimes \mathbb{Z}_{(G)}\langle G \rangle$ are injective and we treat them as identifications.

Thus, for example, when $\mathbb{Z}\langle G\rangle$ is free as a $\mathbb{Z}$–module we will in the natural way regard $\mathbb{Z}\langle G\rangle$ as a subring of $\mathbb{Q}\langle G\rangle$ (this combines all the above identifications:

$$\mathbb{Z}\langle G\rangle \longrightarrow \mathbb{Z}_{(G)} \otimes \mathbb{Z}\langle G\rangle \cong \mathbb{Z}_{(G)}\langle G\rangle \longrightarrow \mathbb{Q}\otimes \mathbb{Z}_{(G)}\langle G\rangle \cong \mathbb{Q}\langle G\rangle).$$

(B) In the setting of Theorem 2.5 we let $\widehat{\sigma}$ denote a specific choice of an element b as in that theorem. With this choice the algebra B is uniquely determined up to (canonical) isomorphism by A, σ, m and c; we denote it by (A, σ, m, c). Thus (A, σ, m, c) is the A–algebra with basis (as a left A–module) $1, \widehat{\sigma}, \cdots, \widehat{\sigma}^{m-1}$ having $\widehat{\sigma}^m = c$ and $\widehat{\sigma}a = \sigma(a)\widehat{\sigma}$ for all $a \in A$.

(C) Suppose A is a $\mathbb{Z}$–algebra which is free as a $\mathbb{Z}$–module. We then use the canonical injective homomorphism $A \longrightarrow \mathbb{Q}\otimes A$ to identify A with a subring of $\mathbb{Q}\otimes A$ and we denote $\mathbb{Q}\otimes A$ by $\mathbb{Q}A$. If in the statement of Theorem 2.5 we have $R = \mathbb{Z}$ and A free as an R–module, then there is a natural isomorphism

$$(\mathbb{Q}A, I\otimes \sigma, m, 1\otimes c) \longrightarrow \mathbb{Q}B$$

(where $I: \mathbb{Q} \longrightarrow \mathbb{Q}$ is the identity map), which we also treat as an identification. Thus in particular A, $\mathbb{Q}A$ and B are all identified with subrings of $\mathbb{Q}B$.

The following lemma is related to an observation of Guralnick and Wiegand [**GW**, p. 564] and will be used in the treatment below of direct products and cyclic extensions of Frobenius complements.

2.7. LEMMA. *Suppose that every nontrivial Sylow subgroup of G has a unique subgroup of prime order. Let P denote the set of prime divisors of $|G|$ and for each $p \in P$ pick a subgroup H_p of G of order p. Then the ideal $(\mathfrak{a}_G)$ of $\mathbb{Z}G$ is generated by $\{\sum_{h\in H_p} \widehat{h} : p \in P\}$.*

PROOF. Suppose $1 \neq h \in G$. Then h^s has prime order p for some integer $s > 0$. By hypothesis (and Sylow's Theorem) $\langle h^s \rangle = bH_p b^{-1}$ for some $b \in G$. Thus the element

$$\sum_{g\in \langle h\rangle} \widehat{g} = \left(\sum_{g\in\langle h^s\rangle} \widehat{g}\right)\left(\sum_{i=0}^{s-1} \widehat{h}^i\right) = \widehat{b}\left(\sum_{g\in H_p} \widehat{g}\right)\widehat{b}^{-1}\left(\sum_{i=0}^{s-1} \widehat{h}^i\right)$$

of $\mathfrak{a}_G$ is in the ideal generated by $\sum_{g\in H_p} \widehat{g}$. $\square$

2.8. CYCLIC EXTENSION LEMMA. *Suppose that $1 \neq H \triangleleft G$ and $c \in G$. Assume that $\mathbb{Z}\langle H\rangle$ is nontrivial and free as a $\mathbb{Z}$-module, that cH generates G/H, that $[G:H]$ divides a power of $|H|$, and that every nontrivial Sylow subgroup of G has a unique subgroup of prime order. Then G is a Frobenius complement, $\mathbb{Z}\langle G\rangle$ is nontrivial and free as a $\mathbb{Z}$-module, and there is a natural isomorphism of $\mathbb{Z}\langle H\rangle$-algebras*

$$\theta: \mathbb{Z}\langle G\rangle \longrightarrow \left(\mathbb{Z}\langle H\rangle, \sigma, [G:H], \overline{c^{[G:H]}}\right)$$

mapping $\overline{c}$ to $\widehat{\sigma}$, where $\sigma \in \mathrm{Aut}\,\mathbb{Z}\langle H\rangle$ is induced by conjugation by c.

The reader will easily verify that the construction of truncated group rings is functorial on the category of groups and injective group homomorphisms. Thus conjugation by c on H does induce an automorphism of $\mathbb{Z}\langle H\rangle$ (as is assumed in the statement of the above theorem) and the inclusion $H \longrightarrow G$ does induce a unitary homomorphism $\mathbb{Z}\langle H\rangle \longrightarrow \mathbb{Z}\langle G\rangle$, making $\mathbb{Z}\langle G\rangle$ a $\mathbb{Z}\langle H\rangle$-algebra.

PROOF. Let $\mathcal{B} = \left(\mathbb{Z}\langle H\rangle, \sigma, [G:H], \overline{c^{[G:H]}}\right)$. It suffices to prove the existence of the isomorphism θ; then by the construction of $\mathcal{B}$, $\mathbb{Z}\langle G\rangle$ is nontrivial and free as a $\mathbb{Z}$–module, and hence G is a Frobenius complement by Lemma 2.2.

We have a well–defined map $\theta_0 : G \longrightarrow \mathcal{B}$ taking hc^i to $\overline{h}\widehat{\sigma}^i$ whenever $h \in H$ and $0 \leq i < [G:H]$. A straightforward computation shows that θ_0 preserves multiplication. Therefore θ_0 induces a ring homomorphism $\theta_1 : \mathbb{Z}G \longrightarrow \mathcal{B}$ which is surjective since $\mathcal{B}$ is generated by $\theta_0(H)$ and $\widehat{\sigma}$. It suffices to show that $\mathfrak{a}_G$ generates the kernel of θ_1. Now for each rational prime p dividing $|G|$, p also divides $|H|$, so G has a subgroup of order p contained in H. Thus by Lemma 2.7 as ideals of $\mathbb{Z}G$ we have $(\mathfrak{a}_G) = (\mathfrak{a}_H) \subset \ker\theta_1$. Now any element A of $\mathbb{Z}G$ can be written in the form $\sum_{i=0}^{[G:H]-1} h_i \widehat{c}^i$ where $h_i \in \mathbb{Z}H$ for all i. (Identify $\mathbb{Z}H$ with a subring of $\mathbb{Z}G$.) Then $\theta_1(A) = \sum_i \overline{h_i}\widehat{\sigma}^i$ where $\overline{h}_i$ denotes the image of h_i in $\mathbb{Z}\langle H\rangle$. Thus, if $A \in \ker\theta_1$ then $\theta_1(h_i) = \overline{h}_i = 0$ for all i, so $h_i \in (\mathfrak{a}_H)$ for all i, and hence $A \in (\mathfrak{a}_G)$. This completes the proof of Lemma 2.8. $\square$

2.9. DIRECT PRODUCT LEMMA. *Let G and H be nontrivial groups of relatively prime orders such that $\mathbb{Z}\langle G\rangle$ and $\mathbb{Z}\langle H\rangle$ are nontrivial and free as $\mathbb{Z}$–modules. Then $\mathbb{Z}\langle G \times H\rangle$ is nontrivial and free as a $\mathbb{Z}$–module, $G \times H$ is a Frobenius complement, and there is an isomorphism $\mathbb{Z}\langle G \times H\rangle \longrightarrow \mathbb{Z}\langle G\rangle \otimes \mathbb{Z}\langle H\rangle$ taking $\overline{(g,h)}$ to $\overline{g} \otimes \overline{h}$ for all $(g,h) \in G \times H$.*

Note that $G \times H$ cannot be a Frobenius complement if G and H do not have relatively prime orders [**Pa**, Theorem 18.1(i), p. 194].

PROOF. By Lemma 2.2 it suffices to prove the existence of the isomorphism. The natural multiplicative map $G \times H \longrightarrow \mathbb{Z}\langle G\rangle \otimes \mathbb{Z}\langle H\rangle$ taking each (g,h) to $\overline{g} \otimes \overline{h}$ induces a homomorphism θ from the integral group ring of $G \times H$ to $\mathbb{Z}\langle G\rangle \otimes \mathbb{Z}\langle H\rangle$. By Lemma 2.7 $(\mathfrak{a}_{G \times H})$ is generated by elements of the forms $\sum_{g \in G_0} \widehat{(g,1)}$ and $\sum_{h \in H_0} \widehat{(1,h)}$ where G_0 and H_0 range over subgroups of G and H, respectively, of prime order. (Recall that $|G|$ and $|H|$ are relatively prime, so every Sylow subgroup of $G \times H$ is a Sylow subgroup of either $G \times 1$ or of $1 \times H$.) θ clearly maps all such elements to 0, so $(\mathfrak{a}_{G \times H}) \subset \ker\theta$. Hence θ induces a homomorphism

$$\theta_1 : \mathbb{Z}\langle G \times H\rangle \longrightarrow \mathbb{Z}\langle G\rangle \otimes \mathbb{Z}\langle H\rangle.$$

On the other hand the injections $G \longrightarrow G \times H$ and $H \longrightarrow G \times H$ induce maps $\mathbb{Z}\langle G\rangle \longrightarrow \mathbb{Z}\langle G \times H\rangle$ and $\mathbb{Z}\langle H\rangle \longrightarrow \mathbb{Z}\langle G \times H\rangle$, and hence a homomorphism

$$\theta_2 : \mathbb{Z}\langle G\rangle \otimes \mathbb{Z}\langle H\rangle \longrightarrow \mathbb{Z}\langle G \times H\rangle$$

with $\theta_2(\overline{g} \otimes \overline{h}) = \overline{(g,h)}$ for all $g \in G$ and $h \in H$. Since θ_2 is clearly the inverse of θ_1, then θ_1 is an isomorphism. $\square$

CHAPTER 3

Groups of Real Quaternions

Denote the division ring of real quaternions by
$$\mathbb{H} = \mathbb{R} + \mathbb{R}\mathbf{i} + \mathbb{R}\mathbf{j} + \mathbb{R}\mathbf{k}$$
where $\mathbb{R}$ denotes the ring of real numbers; we identify $\zeta_n = e^{2\pi i/n}$ with the element $\cos 2\pi/n + \mathbf{i}\sin 2\pi/n$ of $\mathbb{H}$. Let $\tau = (\sqrt{5}+1)/2$ (so $\tau^{-1} = (\sqrt{5}-1)/2$), $\boldsymbol{\alpha} = (-1+\mathbf{i}+\mathbf{j}+\mathbf{k})/2$ and $\boldsymbol{\beta} = (\tau^{-1}+\tau\mathbf{i}+\mathbf{j})/2$.

3.1. REMARK AND EXAMPLES. Consider the following subgroups of $\mathbb{H}^\bullet$, namely, $C_n := \langle \zeta_n \rangle$ ($n \geq 1$), $D_{4n} := \langle \zeta_{2n}, \mathbf{j} \rangle$ ($n \geq 2$), $H_{24} := \langle \mathbf{i}, \mathbf{j}, \boldsymbol{\alpha} \rangle$, $H_{48} := \langle \zeta_8, \mathbf{j}, \boldsymbol{\alpha} \rangle$, and $H_{120} := \langle \mathbf{i}, \mathbf{j}, \boldsymbol{\alpha}, \boldsymbol{\beta} \rangle$. Vignéras [**V**, Theorem 3.7 and Proposition 3.8, p. 17] reports that every finite subgroup of $\mathbb{H}^\bullet$ is isomorphic to one of the groups listed above and that H_{24} and H_{120} are isomorphic to the special linear groups of 2×2 matrices over the fields of order 3 and 5, respectively. By Corollary 3.3 below all of these groups except the trivial group are Frobenius complements. It is easy to see that C_n (for $n > 1$) and D_{4n} (for odd $n > 2$) have core index 1 and that D_{4n} (for even $n \geq 2$) has core index 4. (The core in the latter case is $\langle \zeta_n \rangle$.) Moreover H_{24}, H_{48} and H_{120} have core index 12, 24, and 60, respectively (in each case the core is $\langle -1 \rangle$). In Example 4.3 we construct a Frobenius complement, denoted H_{240}, of core index 120.

We can now state the main theorem of this chapter; equivalent results in the cyclic case go back at least fifty years (see [**LL**, Theorem 2.2]) and include [**BrH**, Claim2, p. 58].

3.2. THEOREM. *Let H be any of the groups C_n ($n \geq 1$), D_{4n} ($n > 1$), H_{24}, H_{48} or H_{120}. Then the inclusion $H \longrightarrow \mathbb{H}$ induces an isomorphism from $\mathbb{Z}\langle H \rangle$ onto $\mathbb{Z}[H]$.*

In the above theorem $\mathbb{Z}[H]$ denotes the subring of $\mathbb{H}$ generated by the set H. We will use a common notation for conjugation in $\mathbb{H}$: for any $\gamma = a + b\mathbf{i} + c\mathbf{j} + d\mathbf{k} \in \mathbb{H}$, we write $\gamma^* = a - b\mathbf{i} - c\mathbf{j} - d\mathbf{k}$ for the conjugate of γ in $\mathbb{H}$. Before proving the theorem we record a corollary of it and Lemma 2.2.

3.3. COROLLARY. *The nontrivial groups listed in Theorem 3.2 are all Frobenius complements.*

In fact all the nontrivial finite subgroups of the group of units of any division ring are Frobenius complements (see [**SW**, Theorem 2.1.2, p. 45] or Proposition 17.1 below).

The next two lemmas will be used in the proof of Theorem 3.2. The first records the results of some routine computations. Note that Part (E) implies that $|\boldsymbol{\alpha}| = 3$.

3.4. LEMMA. *(A) $\boldsymbol{\alpha}\mathbf{i}\boldsymbol{\alpha}^{-1} = \mathbf{k}$, $\boldsymbol{\alpha}\mathbf{j}\boldsymbol{\alpha}^{-1} = \mathbf{i}$, and $\boldsymbol{\alpha}\mathbf{k}\boldsymbol{\alpha}^{-1} = \mathbf{j}$.*

(B) $\beta^4 = \beta^*$ (so $|\beta| = 5$).
(C) $\beta^2 = \alpha\beta + i$.
(D) $(\beta^2 i)^2 = -\alpha\beta i = (\beta^2 i)^*$ (so $|\beta^2 i| = 3$).
(E) If $\gamma = (-1 \pm i \pm j \pm k)/2$, then $\gamma^2 = \gamma^*$ (so $|\gamma| = 3$).

3.5. LEMMA. $1, \beta$ is a basis for $\mathbb{Z}[H_{120}]$ as a $\mathbb{Z}[H_{24}]$-module, and α, i, j, k is a basis for $\mathbb{Z}[H_{24}]$ as a $\mathbb{Z}$-module.

PROOF. Since $|\beta| = 5$ (Lemma 3.4B), $\langle\beta\rangle$ and H_{24} are disjoint subgroups of H_{120} with $|\langle\beta\rangle||H_{24}| = |H_{120}|$, so every element of H_{120} can be uniquely written as a product ab where $a \in H_{24}$ and $b \in \langle\beta\rangle$. Hence Lemma 3.4C implies that the set $\{1, \beta\}$ spans $\mathbb{Z}[H_{120}]$ as a $\mathbb{Z}[H_{24}]$-module. Now suppose $A + B\beta = 0$ where A and B are in $\mathbb{Z}[H_{24}]$ and are not both zero. Then clearly $B \neq 0$ so we can write $\beta = -B^{-1}A$, which is in $\mathbb{Q}[i, j]$ (the ring of rational quaternions is a division subring of $\mathbb{H}$). This is impossible since $\tau \notin \mathbb{Q}$. Hence $1, \beta$ is a basis for $\mathbb{Z}[H_{120}]$ as a $\mathbb{Z}[H_{24}]$-module. The second assertion of the lemma follows easily from [**H**, Lemma 7.4.4, p. 373]. $\square$

Theorem 3.2 will now be proven.

PROOF. By Lemma 2.4 the inclusion $H \longrightarrow \mathbb{H}$ induces a homomorphism $\varphi : \mathbb{Z}H \longrightarrow \mathbb{Z}[H]$ with $\ker\varphi \supset (\mathfrak{a}_H)$. It suffices to prove the reverse inclusion.

Case I: $H = H_{24}$ or H_{48}. Let $\gamma \in \ker\varphi$. We can write $\gamma = \sum_{i=1}^{s} \widehat{h}_i - \sum_{j=1}^{t} \widehat{g}_j$ for some $h_i, g_j \in H$. We may assume by induction that such sums with fewer than $s + t$ terms are in $(\mathfrak{a}_H)$. Note

$$(2) \quad \gamma = \left(\sum_{i=1}^{s} \widehat{h}_i + \sum_{j=1}^{t} \widehat{-g_i}\right) - \left(\sum_{i=1}^{t} \widehat{g}_i \left(\widehat{(-1)} + \widehat{1}\right)\right).$$

The last summation of equation (2) is in $(\mathfrak{a}_H)$. Thus without loss of generality we may assume that $t = 0$, and, further, that $h_1 = 1$ (since $\gamma \in (\mathfrak{a}_H)$ if and only if $\widehat{h}_1^{-1}\gamma \in (\mathfrak{a}_H)$). Hence without loss of generality we may further assume that $h_2 = -1$ or $h_2 = (-1 \pm i \pm j \pm k)/2$; otherwise the real part of $\varphi(\gamma)$ would be of the form $m/2 + n/\sqrt{2}$ where $m, n \in \mathbb{Z}$, $m > 0$ (contradicting that $\varphi(\gamma) = 0$) since the h_i are in H_{48}, whose 48 elements are $\pm 1, \pm i, \pm j, \pm k$, $(\pm 1 \pm i \pm j \pm k)/2$, $(\pm 1 \pm i)/\sqrt{2}$, $(\pm 1 \pm j)/\sqrt{2}$, $(\pm 1 \pm k)\sqrt{2}$, $(\pm i \pm j)/\sqrt{2}$, $(\pm i \pm k)/\sqrt{2}$, $(\pm j \pm k)/\sqrt{2}$ [**C**, p. 372]. If $h_2 = -1$ then $\widehat{h}_1 + \widehat{h}_2 \in \mathfrak{a}_H$ and we are done by induction. On the other hand if $h_2 = (-1 \pm i \pm j \pm k)/2$, then $\widehat{1} + \widehat{h}_2 + \widehat{h}_2^2 \in \mathfrak{a}_H$ by Lemma 3.4E and so

$$\gamma = \left(\widehat{-h_2^2} + \sum_{i=3}^{s} \widehat{h}_i\right) + \left(\widehat{1} + \widehat{h}_2 + \widehat{h}_2^2 - \widehat{h}_2^2(\widehat{1} + \widehat{(-1)})\right).$$

Since the second term is in $(\mathfrak{a}_H)$ and the first has fewer than s elements, we conclude by induction that $\gamma \in (\mathfrak{a}_H)$, as required.

Case II: $H = H_{120}$. First recall from Lemma 3.5 that $1, \beta$ is a basis for $\mathbb{Z}[H_{120}]$ as a $\mathbb{Z}[H_{24}]$-module. Next note that by Lemma 3.4D,

$$\widehat{\beta}^2 - \widehat{\alpha}\widehat{\beta} - \widehat{i} = \left(\widehat{-\alpha\beta i} + \widehat{\beta^2 i} + \widehat{1}\right)\widehat{(-i)} - \left(\widehat{\alpha\beta} + \widehat{-\alpha\beta} + \widehat{i} + \widehat{-i}\right)$$
$$= \left(\widehat{(\beta^2 i)^2} + \widehat{\beta^2 i} + \widehat{1}\right)\widehat{(-i)} - \left(\widehat{\alpha\beta} + \widehat{i}\right)\left(\widehat{1} + \widehat{(-1)}\right) \in (\mathfrak{a}_H).$$

(Be careful here to distinguish between $-\widehat{\gamma}$ and $\widehat{-\gamma}$ for $\gamma \in H$.) Therefore modulo $(\mathfrak{a}_H)$ we have

$$\widehat{\beta}^2 \equiv \widehat{\alpha\beta} + \widehat{\mathbf{i}},$$
$$\widehat{\beta}^3 \equiv \widehat{\alpha\beta^2} + \widehat{\mathbf{i}\beta} = \left(\widehat{\alpha}^2 + \widehat{\mathbf{i}}\right)\widehat{\beta} + \widehat{\alpha\mathbf{i}},$$
$$\widehat{\beta}^4 \equiv \left(\widehat{\alpha}^3 + \widehat{\mathbf{i}\alpha} + \widehat{\alpha\mathbf{i}}\right)\widehat{\beta} + \widehat{\alpha^2\mathbf{i}} + (\widehat{-1}).$$

Now suppose $C \in \ker\varphi$; it suffices to show $C \in (\mathfrak{a}_H)$. C is a sum of terms of the form $n\widehat{\delta}\widehat{\beta}^i$ where $0 \leq i < 5$, $\delta \in H_{24}$ and $n \in \mathbb{Z}$. By the above paragraph C can therefore be written modulo $(\mathfrak{a}_H)$ in the form

$$C \equiv A + B\widehat{\beta}$$

where $A, B \in \mathbb{Z}H_{24}$. Since $\mathfrak{a}_H \subset \ker\varphi$, $0 = \varphi(A) + \varphi(B)\beta$. Since 1, β forms a basis for $\mathbb{Z}[H_{120}]$ as a $\mathbb{Z}[H_{24}]$-module, then $\varphi(A) = \varphi(B) = 0$. Thus by the previous case $A, B \in (\mathfrak{a}_{H_{24}}) \subset (\mathfrak{a}_H)$, so $C \in (\mathfrak{a}_H)$, as required.

Case III: $H = C_n$. The theorem is trivial if $n = 1$. Now suppose n is a prime power p^{t+1} where $t \geq 0$. The canonical surjective homomorphism $\gamma : \mathbb{Z}[x] \longrightarrow \mathbb{Z}H$ with $\gamma(x) = \widehat{\zeta}_n$ carries the cyclotomic polynomial $\Phi_n(x) = \sum_{i=0}^{p-1} X^{ip^t}$ onto $\Phi_n(\widehat{\zeta}_n) = \sum_{i=0}^{p-1} \widehat{\zeta}_p^i$. Thus by Lemma 2.7, γ carries the kernel of the natural homomorphism $\mathbb{Z}[x] \longrightarrow \mathbb{Z}[\zeta_n]$ taking x to ζ_n onto the kernel of the natural homomorphism $\mathbb{Z}H \longrightarrow \mathbb{Z}\langle H\rangle$ (note that $\langle \zeta_p \rangle$ is the unique subgroup of H of prime order). Hence γ induces an isomorphism $\mathbb{Z}[\zeta_n] \longrightarrow \mathbb{Z}\langle H\rangle$ taking ζ_n to $\overline{\zeta}_n$. The inverse of this isomorphism is induced by the inclusion $H \longrightarrow \mathbb{H}$, proving the theorem in this case.

Next suppose $n = mk$ where m and k are relatively prime integers larger than 1. By induction on the number of prime factors of n we may suppose we have isomorphisms $\gamma : \mathbb{Z}\langle C_m\rangle \longrightarrow \mathbb{Z}[\zeta_m]$ and $\delta : \mathbb{Z}\langle C_k\rangle \longrightarrow \mathbb{Z}[\zeta_k]$ taking $\overline{\zeta}_m$ and $\overline{\zeta}_k$ respectively to ζ_m and ζ_k. By the Direct Product Lemma 2.9 we have an isomorphism

$$\mu : \mathbb{Z}\langle C_m \times C_k\rangle \longrightarrow \mathbb{Z}\langle C_m\rangle \otimes \mathbb{Z}\langle C_k\rangle$$

taking $(\overline{\zeta_m, 1})$ to $\overline{\zeta}_m \otimes 1$ and $(\overline{1, \zeta_k})$ to $1 \otimes \overline{\zeta}_k$. We also have natural isomorphisms

$$\eta : \mathbb{Z}\langle C_n\rangle \longrightarrow \mathbb{Z}\langle C_m \times C_k\rangle$$

(induced by the group isomorphism $C_n = \langle \zeta_m, \zeta_k\rangle \longrightarrow C_m \times C_k$ taking ζ_m to $(\zeta_m, 1)$ and ζ_k to $(1, \zeta_k)$) and

$$\rho : \mathbb{Z}[C_m] \otimes \mathbb{Z}[C_k] \longrightarrow \mathbb{Z}[C_n]$$

(taking $\zeta_m \otimes 1$ and $1 \otimes \zeta_k$ to ζ_m and ζ_k, respectively). The composition $\rho(\gamma \otimes \delta)\mu\eta : \mathbb{Z}\langle C_n\rangle \longrightarrow \mathbb{Z}[C_n]$ carries $\overline{\zeta}_k$ and $\overline{\zeta}_m$ to ζ_k and ζ_m, and hence is the promised isomorphism induced by the inclusion $H \longrightarrow \mathbb{Z}[H]$.

Case IV: $H = D_{4n}$, $n \geq 2$. By the Cyclic Extension Lemma 2.8 we have an isomorphism

$$\mathbb{Z}\langle D_{4n}\rangle \longrightarrow (\mathbb{Z}\langle C_{2n}\rangle, \sigma, 2, -1)$$

taking $\overline{\zeta}_{2n}$ (in $\mathbb{Z}\langle D_{4n}\rangle$) to $\overline{\zeta}_{2n}$ (in $\mathbb{Z}\langle C_{2n}\rangle$) and $\overline{\mathbf{j}}$ to $\widehat{\sigma}$, where σ is the automorphism of $\mathbb{Z}\langle C_{2n}\rangle$ induced by conjugation by $\mathbf{j}$ (so $\sigma(\overline{\zeta}_{2n}) = \overline{\mathbf{j}\zeta_{2n}\mathbf{j}^{-1}} = \overline{\zeta}_{2n}^{-1}$). By the

previous case we have an isomorphism
$$(\mathbb{Z}\langle C_{2n}\rangle, \sigma, 2, -1) \longrightarrow (\mathbb{Z}[\zeta_{2n}], \tau, 2, -1)$$
where $\tau(\zeta_{2n}) = \zeta_{2n}^{-1}$. Composing these isomorphisms with the obvious isomorphism
$$(\mathbb{Z}[\zeta_{2n}], \tau, 2, -1) \longrightarrow \mathbb{Z}[D_{4n}]$$
(map ζ_{2n} to itself and $\hat{\tau}$ to $\mathbf{j}$) gives the required map $\mathbb{Z}\langle H\rangle \longrightarrow \mathbb{Z}[H]$. $\square$

The next lemma will be needed in several proofs below, beginning with the proof of the Classification Theorem 1.4 in the next chapter.

3.6. LEMMA. *(A) There is a unique automorphism ψ_0 of H_{120} mapping $\mathbf{j}$ to $\mathbf{k}$ and fixing $\mathbf{i}$. It maps $\boldsymbol{\alpha}$ to $(\boldsymbol{\alpha}\mathbf{i})^{-1}$ and $\boldsymbol{\beta}$ to $(\boldsymbol{\beta}\mathbf{i})^{-1}$.*
(B) There is a unique automorphism of $\mathbb{Z}[H_{120}]$ mapping $\mathbf{j}$ to $\mathbf{k}$ and fixing $\mathbf{i}$. It extends the map ψ_0.
(C) There is a unique automorphism of $\mathbb{Q}[\mathbf{i}, \mathbf{j}, \sqrt{5}]$ extending ψ_0. It maps $\sqrt{5}$ to $-\sqrt{5}$.

PROOF. Conjugation by ζ_8 fixes $\mathbf{i} = \zeta_8^2$ and maps $\mathbf{j}$ to
$$\zeta_8 \mathbf{j} \zeta_8^{-1} = \zeta_8 (\mathbf{j}\zeta_8^{-1}\mathbf{j}^{-1})\mathbf{j} = \zeta_8^2 \mathbf{j} = \mathbf{ij} = \mathbf{k}.$$
Hence there is an automorphism ψ_1 of $\mathbb{Q}[\mathbf{i},\mathbf{j}]$ of order 4 fixing $\mathbf{i}$ and taking $\mathbf{j}$ to $\mathbf{k}$. Let ψ_2 be the automorphism of $\mathbb{Q}[\sqrt{5}]$ of order 2 taking $\sqrt{5}$ to $-\sqrt{5}$. Then $\psi_1 \otimes \psi_2$ induces an automorphism ψ of order 4 of
$$\mathbb{Q}[H_{120}] = \mathbb{Q}[\mathbf{i}, \mathbf{j}, \sqrt{5}] (\cong \mathbb{Q}[\mathbf{i}, \mathbf{j}] \otimes \mathbb{Q}[\sqrt{5}])$$
taking $\sqrt{5}$ to $-\sqrt{5}$, $\mathbf{j}$ to $\mathbf{k}$ and fixing $\mathbf{i}$. Then $\psi(\mathbf{k}) = \psi(\mathbf{ij}) = \mathbf{ik} = -\mathbf{j}$ and $\psi(\tau) = -\tau^{-1}$, so $\psi(\boldsymbol{\alpha}) = (-1 + \mathbf{i} - \mathbf{j} + \mathbf{k})/2 = (\boldsymbol{\alpha}\mathbf{i})^{-1}$ and
$$\psi(\boldsymbol{\beta}) = (-\tau - \tau^{-1}\mathbf{i} + \mathbf{k})/2 = (\boldsymbol{\beta}\mathbf{i})^{-1}.$$
The automorphism ψ therefore restricts to an automorphism ψ_0 of H_{120} and to an automorphism of $\mathbb{Z}[H_{120}]$ fixing $\mathbf{i}$ and mapping $\mathbf{j}$ to $\mathbf{k}$. (The restrictions must be automorphisms since ψ has order 4.) Suppose now that φ is another automorphism of $\mathbb{Z}[H_{120}]$ fixing $\mathbf{i}$ and mapping $\mathbf{j}$ to $\mathbf{k}$. Then φ extends to an automorphism φ_1 of $\mathbb{Q}[H_{120}](\cong \mathbb{Q} \otimes \mathbb{Z}[H_{120}])$. The center of $\mathbb{Q}[H_{120}] = \mathbb{Q}[\sqrt{5}][\mathbf{i},\mathbf{j}]$ is $\mathbb{Q}[\sqrt{5}]$, so φ_1 restricts to an automorphism of $\mathbb{Q}[\sqrt{5}]$. If φ_1 fixes $\sqrt{5}$, then $\varphi_1(\tau) = \tau$, so $\varphi(\boldsymbol{\beta}) = (\tau^{-1} + \tau\mathbf{i} + \mathbf{k})/2$ and hence
$$-\frac{1}{2} = [(\mathbf{k} - \mathbf{j})/2]^2 = (\varphi(\boldsymbol{\beta}) - \boldsymbol{\beta})^2 \in \mathbb{Z}[H_{120}].$$
But this contradicts the fact that $\mathbb{Z}[H_{120}]$ as a $\mathbb{Z}$–module is finitely generated. Hence $\varphi_1(\sqrt{5}) = -\sqrt{5}$, so $\varphi_1 = \psi$. Thus $\varphi = \psi|\mathbb{Z}[H_{120}]$, so $\psi|\mathbb{Z}[H_{120}]$ is the unique automorphism of $\mathbb{Z}[H_{120}]$ fixing $\mathbf{i}$ and mapping $\mathbf{j}$ to $\mathbf{k}$. Now suppose φ_0 is a group automorphism of H_{120} fixing $\mathbf{i}$ and mapping $\mathbf{j}$ to $\mathbf{k}$. Then φ_0 induces an automorphism of $\mathbb{Z}\langle H_{120}\rangle$ and hence by Theorem 3.2 it extends to an automorphism of $\mathbb{Z}[H_{120}]$ fixing $\mathbf{i}$ and mapping $\mathbf{j}$ to $\mathbf{k}$. Then by the uniqueness of $\psi|\mathbb{Z}[H_{120}]$ we must have $\varphi_0 = \psi_0$. Finally suppose φ_2 is any extension of ψ_0 to an automorphism of $\mathbb{Q}[H_{120}]$. If $\varphi_2(\sqrt{5}) = -\sqrt{5}$, then $\varphi_2 = \psi$ and we are finished. If not we must have $\varphi_2(\sqrt{5}) = \sqrt{5}$, so
$$\varphi_2(\boldsymbol{\beta}) = (\tau^{-1} + \tau\mathbf{i} + \mathbf{k})/2 \neq \psi_0(\boldsymbol{\beta}),$$
contradicting that φ_2 extends ψ_0. Thus $\varphi_2 = \psi$. $\square$

CHAPTER 4

Proof of the Classification Theorem

We now prove Theorem 1.4. The proof will use some properties of generalized quaternion groups which are collected in the lemma below. Recall that the Sylow 2–subgroups of Frobenius complements are either cyclic (in which case the Frobenius complement is a $\mathbb{Z}$–group) or generalized quaternion groups. The generalized quaternion group of order $2^{n+1} \geq 8$ admits generators u and v with $|v| = 2^n$, $u^2 = v^{2^{n-1}}$ and $uvu^{-1} = v^{-1}$. The proof of the next lemma uses this notation.

4.1. LEMMA. *Let H be a generalized quaternion group of order 2^{n+1} and let $z \in H$ have order 2^n.*
  (A) *If $w \in H \backslash \langle z \rangle$, then $H = \langle w, z \rangle$, $|w| = 4$, and $wzw^{-1} = z^{-1}$.*
  (B) $H/\langle z^2 \rangle \cong V_4$.
  (C) *Suppose $0 \leq s \leq n$, $s \neq 2$. Then $\langle z^{2^{n-s}} \rangle$ is the unique cyclic subgroup of H of order 2^s. If $s = 2$ and $n > 2$, then it is the unique cyclic normal subgroup of H of order 2^s.*

PROOF. Write $H = \langle u, v \rangle$ as in the paragraph above the statement of the lemma. The assertions of the lemma are easily verified if $n = 2$ (*i.e.*, H is a quaternion group), so suppose $n > 2$. Since $v^{2^{n-1}} = u^2$, then every element of H has the form v^i or uv^i. But for all i, $(uv^i)^2 = u^2$, so uv^i clearly has order 4. Hence z must be a power of v. All the assertions of (A) and (B) are now easily checked. The fact that all elements of H outside of $\langle v \rangle$ have order 4 also makes clear that H has only one cyclic subgroup of order 2^s whenever $0 \leq s \leq n$ except when $s = 2$, and this subgroup must be $\langle z^{2^{n-s}} \rangle$. Now suppose $s = 2$. Then $\langle z^{2^{n-s}} \rangle$ is normal and cyclic of order 4. Any other such subgroup would have the form $\langle uv^i \rangle$, and hence would contain the element

$$v(uv^i)v^{-1}(uv^i)^3 = v^2.$$

Since v^2 has order $2^{n-1} \geq 4$ we must have $\langle uv^i \rangle = \langle v^2 \rangle$. Hence $u \in \langle v \rangle$, a contradiction. Thus $\langle z^{2^{n-s}} \rangle$ is the only normal cyclic subgroup of H of order 4. □

4.2. LEMMA. *Let G be a Frobenius complement with a subgroup of index 2 which is an internal direct product of H_{120} and a subgroup J. Then there exists an element u of G of order 8 with $u^2 = \boldsymbol{i}$, $u\boldsymbol{j}u^{-1} = \boldsymbol{k}$, $u\boldsymbol{k}u^{-1} = -\boldsymbol{j}$, $u\alpha u^{-1} = (\alpha \boldsymbol{i})^{-1}$ and $u\beta u^{-1} = (\beta \boldsymbol{i})^{-1}$.*

PROOF. The order of J is relatively prime to 30 [**Pa**, Theorem 18.1(i), p. 193-194]. Thus H_{120} is a characteristic subgroup of JH_{120} and any Sylow 2–subgroup of G must be a generalized quaternion group of order 16. G has such a subgroup S containing $\langle \boldsymbol{i}, \boldsymbol{j} \rangle$, say with $u \in S$ of order 8. Then u^2 is an element of order 4 in $\langle \boldsymbol{i}, \boldsymbol{j} \rangle$. By Lemma 3.4A we may assume without loss of generality that u (and S) are chosen with $u^2 = \pm \boldsymbol{i}$ and indeed with $u^2 = \boldsymbol{i}$ (if necessary, replace u by $\boldsymbol{i}u$).

4. PROOF OF THE CLASSIFICATION THEOREM

Then by Lemma 4.1A

$$uju^{-1} = u(\mathbf{j}u\mathbf{j}^{-1})^{-1}\mathbf{j} = u^2\mathbf{j} = \mathbf{ij} = \mathbf{k}$$

and so $u\mathbf{k}u^{-1} = u\mathbf{ij}u^{-1} = \mathbf{ik} = -\mathbf{j}$. Finally, applying Lemma 3.6A to conjugation by u, we have $u\boldsymbol{\alpha}u^{-1} = (\boldsymbol{\alpha}\mathbf{i})^{-1}$ and $u\boldsymbol{\beta}u^{-1} = (\boldsymbol{\beta}\mathbf{i})^{-1}$. □

The above lemma shows that there is at most one (up to isomorphism) Frobenius complement having H_{120} as a subgroup of index 2. Such a group is constructed in Example 4.3 below.

We now give the proof of the Classification Theorem 1.4.

PROOF. Recall that a $\mathbb{Z}$–group is a group all of whose Sylow subgroups are cyclic, so a Frobenius complement is a $\mathbb{Z}$–group if and only if its Sylow 2–subgroup is cyclic [**Pa**, Theorem 18.1(iv), p. 194]. Hence a Frobenius complement is a $\mathbb{Z}$–group if it has a subgroup of odd index which is a $\mathbb{Z}$–group.

Now let G be any Frobenius complement. We begin by showing the existence of a core N of G, i.e., a normal $\mathbb{Z}$–subgroup of G such that G/N is isomorphic to one of the groups on the list (1) of Chapter 1. We first suppose G is solvable. Then G has a normal subgroup N_0 which is a $\mathbb{Z}$–group of maximal possible order such that G/N_0 is isomorphic to a subgroup of S_4 [**Pa**, Theorem 18.2 (Zassenhaus), p. 196]. Therefore $[G : N_0] \neq 3$, since otherwise by the previous paragraph G is a $\mathbb{Z}$–group and so $|N_0|$ was not maximal. Similarly, $[G : N_0] \neq 6$ (otherwise N_0 would be a subgroup of index 3 of a subgroup of G of index 2, since every group of order 6 has a subgroup of order 3). If $[G : N_0] = 1$, 12, or 24, then G/N_0 is isomorphic to 1, A_4, or S_4 and we can simply set $N = N_0$. Suppose this is not the case; then by Lagrange's theorem $[G : N_0]$ is a proper power of 2: 2, 4 or 8. We may write $N_0 = \langle x, y \rangle$ where $|x| = n$, $|y| = m$ and $xy = y^r x$ for some positive integers r, n, m satisfying $(m, nr(r-1)) = 1$ [**Pa**, Proposition 12.11, p. 106],[**S**, 12.6.17, p. 356]. Then m is clearly odd. Recall that 2_n denotes the largest power of 2 dividing n and $n//2$ denotes the largest odd factor of n (cf. Chapter 1). Then $N_1 := \langle x^{2_n}, y \rangle$ is clearly a normal subgroup of G of odd order $|G|//2$. Let H be a Sylow 2–subgroup of G; H is a generalized quaternion group since G is not a $\mathbb{Z}$–group. The natural homomorphism $H \longrightarrow G/N_1$ is injective, and hence is an isomorphism. By Lemma 4.1B there is a normal cyclic subgroup H_1 of H with $H/H_1 \cong V_4$. Hence by the Noether isomorphism theorems G has a normal subgroup N containing N_1 with $N/N_1 \cong H_1$ and $G/N \cong V_4$. N is a $\mathbb{Z}$–group since its Sylow 2–subgroups are isomorphic to the cyclic group H_1. This completes the proof of the existence of the group N of Theorem 1.4 in the case that G is solvable.

Now suppose G is not solvable. Then without loss of generality we may assume that G has a subgroup of index at most two which is an internal direct product of a $\mathbb{Z}$–group M of order relatively prime to 30 and the group H_{120} of the previous chapter (apply [**V**, Proposition 3.8, p. 17] and [**Pa**, Theorem 18.6 (Zassenhaus), p. 204]). $\mathbb{Z}^\bullet = \{\pm 1\}$ is the unique subgroup of H_{120} of order 2; it is in the center of G [**Pa**, Theorem 18.1(iii), p. 194] and is normal in G. M is normal in G because MH_{120} is normal in G and M is a characteristic subgroup of MH_{120} (it is the unique subgroup of MH_{120} of order $|M|$). Thus $N := \mathbb{Z}^\bullet M$ is a normal $\mathbb{Z}$–subgroup of G. If $G = MH_{120}$, then $G/N \cong H_{120}/\mathbb{Z}^\bullet \cong A_5$ [**Pa**, Lemma 13.6 and Proposition 13.7, pp. 120–122]. Now suppose $G \neq MH_{120}$. Then

$$|G/N| = [G : MH_{120}][MH_{120} : MZ^\bullet] = 2|A_5| = |S_5|.$$

We can therefore show that $G/N \cong S_5$ by proving that G/N is generated by elements satisfying defining relations for S_5.

By Lemma 4.2 G has an element u with $u^2 = \mathbf{i}$, $u\mathbf{j}u^{-1} = \mathbf{k}$, $u\mathbf{k}u^{-1} = -\mathbf{j}$, $u\boldsymbol{\alpha}u^{-1} = (\boldsymbol{\alpha}\mathbf{i})^{-1}$ and $u\boldsymbol{\beta}u^{-1} = (\boldsymbol{\beta}\mathbf{i})^{-1}$. Let $s_2 = \boldsymbol{\beta}u^{-1}N$ and $s_5 = \boldsymbol{\beta}N$ in G/N. In order to show that $G/N \cong S_5$ it then suffices to prove that $G/N = \langle s_2, s_5 \rangle$ and

$$(3) \qquad s_2^2 = s_5^2 = (s_5 s_2)^4 = \left(s_5^{-1} s_2 s_5 s_2\right)^3 = 1$$

[**B**, Section 6, p. 125]. Now $\langle s_2, s_5 \rangle$ contains $\boldsymbol{\beta}N$, uN, and $\mathbf{i}N = u^2 N$ and therefore also contains $\boldsymbol{\alpha}N$ (Lemma 3.4D) and hence both $\mathbf{j}N$ and $\mathbf{k}N$ (Lemma 3.4A). Thus $\langle s_2, s_5 \rangle$ contains and hence equals

$$\langle u \rangle (H_{120} M)/N = G/N.$$

Next, $s_5^5 = 1$ by Lemma 3.4B, and

$$s_2^2 = u^{-1}(u\boldsymbol{\beta}u^{-1})\boldsymbol{\beta}u^{-1}N = u^{-1}\mathbf{i}^{-1}\boldsymbol{\beta}^{-1}\boldsymbol{\beta}u^{-1}N = -1N = 1.$$

Therefore

$$\begin{aligned}(s_5 s_2)^4 &= \boldsymbol{\beta}N s_2^2 uN s_2^2 uN s_2^2 uN s_2^2 u\boldsymbol{\beta}^{-1}N \\ &= \boldsymbol{\beta}u^4 \boldsymbol{\beta}^{-1}N = 1.\end{aligned}$$

Also since $|\boldsymbol{\beta}^2 \mathbf{i}| = 3$ (Lemma 3.4D), we have

$$\begin{aligned}\left(s_5^{-1} s_2 s_5 s_2\right)^3 &= u^{-1}\boldsymbol{\beta}^2 u^{-2} \boldsymbol{\beta}^2 u^{-2} \boldsymbol{\beta}^2 u^{-2} uN \\ &= u^{-1}(\boldsymbol{\beta}^2 \mathbf{i})^3(-u)N = -1N = 1.\end{aligned}$$

This completes the proof of the relations in the list (3). Thus $G/N \cong S_5$. This completes the proof of the existence part of the Classification Theorem 1.4. We next prove uniqueness.

Let us suppose that M and N are both normal subgroups of G which are $\mathbb{Z}$-groups such that G/N and G/M are both isomorphic to groups on the list (1). We prove $M = N$. We may suppose without loss of generality that $[G : N] \leq [G : M]$ and hence that $[G : N]$ divides $[G : M]$. (Note that if G is solvable, then $[G : N]$ and $[G : M]$ are both in $\{1, 4, 12, 24\}$, and otherwise they are both in $\{60, 120\}$.) Thus $|M|$ divides $|N|$.

We claim $M \subset N$. Without loss of generality (for proving this assertion) $G \neq N$. Hence $G \neq M$. Hence $[G : N]$ and $[G : M]$ are divisible by 4. For each prime divisor p of $|G|$ we can find Sylow p–subgroups M_p, M_p^*, N_p, and N_p^* of M, G, N, and G, respectively, with $M_p \subset M_p^*$ and $N_p \subset N_p^*$. Consider any such prime p. Since $M_p = M \cap M_p^*$, then $M_p \triangleleft M_p^*$. Similarly $N_p \triangleleft N_p^*$, and if $p = 2$ then 4 divides $[N_p^* : N_p]$. Since N_p is cyclic it has a cyclic subgroup N_0 of order $|M_p|$. N_0 is a characteristic subgroup of N_p and $N_p \triangleleft N_p^*$, so $N_0 \triangleleft N_p^*$. Indeed, N_0 is the unique normal cyclic subgroup of N_p^* of order $|M_p|$ (this is trivial if N_p^* is cyclic; otherwise apply Lemma 4.1C after observing that 4 divides $[N_p^* : N_0]$). Now Sylow p–subgroups of G are conjugate, so there exists $x \in G$ with $xM_p^* x^{-1} = N_p^*$. Since $M_p \triangleleft M_p^*$, then $xM_p x^{-1} \triangleleft xM_p^* x^{-1} = N_p^*$. Hence $xM_p x^{-1} = N_0$. Thus $M_p = x^{-1} N_0 x \subset x^{-1} N x = N$. But then, as claimed, $M = \langle \bigcup M_p \rangle \subset N$, where the union is taken over all prime divisors p of $|M|$.

We finish the proof of Theorem 1.4 by proving that $N = M$. First suppose G is nonsolvable. Then G/N and G/M must be isomorphic to either A_5 or S_5. Now G/N is a homomorphic image of G/M but A_5 is not a homomorphic image of

S_5 (otherwise S_5 has a normal subgroup of order 2 and hence a central element of order 2, which is false). Thus G/N and G/M must be isomorphic to each other, so $|N| = |M|$, and hence $N = M$. Next suppose G is solvable, so G/N and G/M are isomorphic to 1, V_4, A_4 or S_4. Again, suppose G/N is not isomorphic to G/M since otherwise $|M| = |N|$, so $N = M$. However G/N is again a homomorphic image of G/M. But of the groups 1, V_4, A_4 and S_4 only 1 is a homomorphic image of any of the others. Thus $G = N$, and so G is a $\mathbb{Z}$-group. Hence the Sylow 2-subgroups of G, and therefore of G/M also, are all cyclic. But neither V_4, A_4 nor S_4 has a cyclic 2-Sylow subgroup. Thus G/M is trivial, so $M = G = N$ in this case. This completes the proof of the Classification Theorem. □

The following construction of a Frobenius complement with core index 120 will preview techniques which will be used repeatedly in the following chapters.

4.3. EXAMPLE. Let ψ denote the automorphism of $\mathbb{Z}[H_{120}]$ of Lemma 3.6B. Recall that $\psi(\mathbf{i}) = \mathbf{i}$ and note that ψ^2 is just conjugation by $\mathbf{i}$. Thus the ring $R := (\mathbb{Z}[H_{120}], \psi, 2, \mathbf{i})$ is well-defined and free as a $\mathbb{Z}$-module (cf. Remark 2.6 and Lemma 3.5). Let H_{240} denote the subgroup of $R^\bullet$ generated by H_{120} and $\widehat{\psi}$. Then H_{240} satisfies the conditions of the Cyclic Extension Lemma 2.8 (with $H = H_{120}$ and $c = \widehat{\psi}$; note that the Sylow subgroup $\langle \widehat{\psi}, \mathbf{j} \rangle$ is a generalized quaternion group of order 16). Thus $\mathbb{Z}\langle H_{240} \rangle \cong R$ is nontrivial and torsion-free as a $\mathbb{Z}$-module and H_{240} is a Frobenius complement. Since H_{240} contains H_{120}, it is not solvable, so it must have core index 60 or 120. Just suppose the core index is 60. Since $|H_{240}| = 240 = 4|A_5|$, then the core would be a cyclic group of order 4. By Sylow's Theorem some conjugate of the core, and hence the core itself, is a normal cyclic subgroup of $\langle \widehat{\psi}, \mathbf{j} \rangle$ of order 4. But by Lemma 4.1C the only such subgroup is $\langle \widehat{\psi}^2 \rangle = \langle \mathbf{i} \rangle$. This is impossible, however, since $\langle \mathbf{i} \rangle$ is not even a normal subgroup of H_{120} (e.g. see Lemma 3.4A). Thus H_{240} must have core index 120. It follows that the core of H_{240} is $\langle -1 \rangle$.

We end this chapter by sketching an alternative approach to classifying Frobenius complements into six types (cf. Theorem 1.4).

4.4. REMARK. If G has core index larger than 1, then the core C admits a unique subgroup C_0 of index 2 and the factor group G/C_0 is isomorphic to either D_8, H_{24}, H_{48}, H_{120} or H_{240} depending on whether G has core index 4, 12, 24, 60 or 120, respectively. This fact could be used in place of the Classification Theorem 1.4. Our explicit use of the groups "H_i" (e.g. in Theorems 7.11 and 9.1) recommends such an approach. On the other hand, the groups V_4, A_4, S_4, A_5 and S_5 used in Theorem 1.4 are simpler and more familiar than the groups D_8, H_{24}, H_{48}, H_{120} and H_{240}.

CHAPTER 5

Frobenius complements with core index 1

By a *1-complement* we mean a Frobenius complement which is a $\mathbb{Z}$-group; this usage will be generalized in Definition 6.1 below. We study here the isomorphism classes and integral truncated group rings of 1-complements; this will require a modest elaboration of the usual structure theory of $\mathbb{Z}$-groups [**Pa**, Proposition 12.11, p. 106]. Recall that if $0 < n \in \mathbb{Z}$, then n_0 denotes the largest square-free integer dividing n.

5.1. DEFINITION. (A) A *Frobenius triple* is an ordered triple (m, n, T) where
  (A1) m and n are relatively prime positive integers;
  (A2) the order of T divides n/n_0; and
  (A3) T is a cyclic subgroup of $\mathbb{Z}_m^\bullet$ with $(-1+T) \cap \mathbb{Z}_m^\bullet$ nonempty.
(B) We call the Frobenius triple (m, n, T) *proper* if $n > 1$.
(C) Suppose G is a $\mathbb{Z}$-group. The *invariant* of G is the triple
$$(|G'|, [G:G'], I)$$
where I is the image of the composition of functions
$$G \longrightarrow \operatorname{Aut} G' \longrightarrow Z_{|G'|}^\bullet.$$
The left hand map above is induced by conjugation by elements of G and the right hand map is the natural isomorphism (recall that G' is cyclic [**S**, 12.6.17, p. 356]).
(D) If G has invariant $(m, n, \langle r \rangle)$ for some $r \in \mathbb{Z}_m^\bullet$, then an *r-sequence* for G is a sequence x, y of elements of G with $|x| = n$, $|y| = m$, and $xyx^{-1} = y^r$.

The only Frobenius triple which is not proper is $(1, 1, \mathbb{Z}_1^\bullet)$.
The next theorem assembles the main results of this chapter.

5.2. THEOREM. *(A) The map assigning to each 1-complement its invariant induces a bijection from the set of isomorphism classes of 1-complements to the set of proper Frobenius triples.*
(B) If G is a 1-complement with invariant $(m, n, \langle r \rangle)$, then G has an r-sequence.
(C) Say $(m, n, \langle r \rangle)$ is a proper Frobenius triple and $t = |r|$. Then there is an automorphism σ of $\mathbb{Z}[\zeta_{mn/t}]$ fixing $\zeta_{n/t}$ and with $\sigma(\zeta_m) = \zeta_m^r$. The subgroup $G := \langle \zeta_m, \widehat{\sigma} \rangle$ of the group of units of the ring
$$\mathcal{A} := (\mathbb{Z}[\zeta_{mn/t}], \sigma, t, \zeta_{n/t})$$
is a 1-complement with invariant $(m, n, \langle r \rangle)$. Moreover, the inclusion $G \longrightarrow \mathcal{A}$ induces an isomorphism $\mathbb{Z}\langle G \rangle \longrightarrow \mathcal{A}$.
(D) Let G be as in part (C). Then $\mathbb{Z}\langle G \rangle$ is a free $\mathbb{Z}$-module of rank $\phi(mn)$, and $\mathbb{Q}\langle G \rangle$ is a simple algebra with degree t and with rational dimension $[\mathbb{Q}\langle G \rangle : \mathbb{Q}] = \phi(mn)$; the isomorphism $\mathbb{Q}\langle G \rangle \longrightarrow \mathbb{Q}\mathcal{A}$ maps the center of $\mathbb{Q}\langle G \rangle$ onto $\mathbb{Q}[\zeta_{mn/t}]^\sigma$.

The "σ" in part (D) is actually the canonical extension of the σ of part (C) to $\mathbb{Q}[\zeta_{mn/t}]$. The bijection of part (A) easily generalizes from 1–complements to arbitrary $\mathbb{Z}$–groups. The focus in (A) on the group T rather than on the generator r (as in [**Pa**]) is because T, but not r, is an isomorphism invariant of G. The index of the simple algebra $\mathbb{Q}\langle G\rangle$ is computed in Theorem 17.4 below.

We begin the proof of Theorem 5.2 by collecting some simple observations about Frobenius triples.

5.3. LEMMA. *Let m and n denote relatively prime positive integers. Let $r \in \mathbb{Z}_m^\bullet$ and set $t = |r|$ and $T = \langle r \rangle$.*

(A) *(m, n, T) satisfies condition (A2) of Definition 5.1 if and only if $(m, r-1) = 1$; also, (m, n, T) satisfies condition (A3) of Definition 5.1 if and only if $r^{n/n_0} = 1$.*

(B) *Suppose (m, n, T) is a Frobenius triple. Then:*
   (B1) *m is odd, and either t is even or $3 \nmid m$;*
   (B2) *if $n = 2$ then $m = t = 1$;*
   (B3) *if $n \equiv 2 \pmod 4$, then t is odd and $(m, n/2, T)$ is a Frobenius triple.*

PROOF. Note that if $-1 + r^j \in \mathbb{Z}_m^\bullet$, then $1 = ((r-1)(1 + r + \cdots + r^{j-1}), m)$, so $(m, r-1) = 1$. We leave the rest of the proof of part (A) to the reader. The equation $(m, r) = (m, r-1) = 1$ implies m is odd. It also says that if 3 divides m, then $r \equiv 2 \pmod 3$ so $r^2 \equiv 1 \pmod 3$. But if 3 divides m, then $r^t \equiv 1 \pmod 3$, so t must be even. (If t were odd then $r \equiv 1 \pmod 3$, contradicting that $(m, r-1) = 1$.) The last two parts of (B) are easily verified. □

We now give the proof of Theorem 5.2.

PROOF. Let G be a 1–complement. Then by Theorem 12.11 of [**Pa**, p.106] there exists $x, y \in G$, positive integers m and n, and $r \in \mathbb{Z}_m^\bullet$ with $r^n = 1$ and $(m, nr(r-1)) = 1$ such that the equations

$$x^n = y^m = 1, \quad xy = y^r x$$

give a presentation of G. Moreover by Theorem 18.2 of [**Pa**, p. 196] (or, more pecisely, by the first paragraph of its proof) $r^{n/n_0} \equiv 1 \pmod m$, so $(m, n, \langle r \rangle)$ is a Frobenius triple (cf. Lemma 5.3). Since $G' = \langle y \rangle$ (note $xyx^{-1}y^{-1} = y^{r-1}$ generates $\langle y \rangle$), then $m = |G'|$ and $n = [G : G']$. The image of the map $G \longrightarrow \operatorname{Aut} G'$ is generated by the image of x and this maps to r in $\mathbb{Z}_m^\bullet$. Thus the invariant of G is $(m, n, \langle r \rangle)$, and it is a proper Frobenius triple.

Suppose s is any generator of $\langle r \rangle$. Let $t = |r|$, so t divides n. Then $s = r^i$ for some $i \in \mathbb{Z}$ such that $(t, i) = 1$. There exists $j \in \mathbb{Z}$ with $j \equiv i \pmod {t_n}$ and $j \equiv 1 \pmod {n/t_n}$ where t_n as usual denotes the largest divisor of n dividing some power of t. Then $(j, n) = 1$, so $|x^j| = n$. Then x^j, y is an s–sequence for G, proving part (B).

Next suppose G_1 is another 1–complement with the same invariant $(m, n, \langle r \rangle)$ as G. By (B), G_1 has an r–sequence x_1, y_1. But then $G \cong G_1$ since the two groups have equivalent presentations in terms of the generating sets x, y and x_1, y_1.

We next prove part (C); note that this will complete the proof of part (A). Thus let $(m, n, \langle r \rangle)$ be any proper Frobenius triple. The existence of the automorphism σ of (C) is routine since $(m, n) = (m, r) = 1$. Note σ has order $t = |r|$. Thus the algebra $\mathcal{A}$ of part (C) is well–defined (cf. Theorem 2.5 and Remark 2.6B). Let $H = \langle \zeta_{mn/t} \rangle$ and $G = \langle \zeta_m, \widehat{\sigma} \rangle$. Then $|\zeta_m| = m$ and since $\widehat{\sigma}^t = \zeta_{n/t}$, then $|\widehat{\sigma}| = n$.

(We clearly have $\widehat{\sigma}^n = (\zeta_{n/t})^{n/t} = 1$. On the other hand if $1 = \widehat{\sigma}^{tq+\rho}$ where $0 \le \rho < t$, then $1 = (\zeta_{n/t})^q \widehat{\sigma}^\rho$, so by the definition of $\mathcal{A}$, n/t divides q and $\rho = 0$, so n divides $tq+\rho$.) Moreover $\widehat{\sigma}\zeta_m\widehat{\sigma}^{-1} = \sigma(\zeta_m) = \zeta_m^r$. Hence G is clearly a $\mathbb{Z}$-group of order mn. By Theorem 3.2 the inclusion $H \longrightarrow \mathbb{Z}[\zeta_{mn/t}]$ induces an isomorphism $\mathbb{Z}\langle H\rangle \longrightarrow \mathbb{Z}[\zeta_{mn/t}]$. Hence by Corollary 3.3 H is a Frobenius complement. Since $\mathbb{Z}[\zeta_{mn/t}]$ is torsion-free and finitely generated as a $\mathbb{Z}$-module, then $\mathbb{Z}\langle H\rangle$ is free and of rank $\phi(mn/t)$. Condition (A3) of Definition 5.1 says that $[G:H] = t$ divides a power of $|H| = mn/t$. Hence by the Cyclic Extension Lemma 2.8 G is a Frobenius complement, and hence a 1–complement with invariant $(m, n, \langle r\rangle)$, and the inclusion $G \longrightarrow \mathcal{A}$ induces an isomorphism $\mathbb{Z}\langle G\rangle \longrightarrow \mathcal{A}$. Thus in particular $\mathbb{Z}\langle G\rangle$ is free of rank $t\phi(mn/t) = \phi(mn)$ (use Definition 5.1(A3) to apply the usual computation of the Euler ϕ-function [**NZ**, Section 2.4, pp. 48-49]). Hence $\mathbb{Q}\langle G\rangle$ has rational dimension $\phi(mn)$. Since $\mathbb{Q}\langle G\rangle \cong (\mathbb{Q}[\zeta_{mn/t}], \sigma, t, \zeta_{n/t}) = \mathbb{Q}\mathcal{A}$ (cf. Remark 2.6), by Theorem 2.5 $\mathbb{Q}\langle G\rangle$ is simple with center isomorphic to $\mathbb{Q}[\zeta_{mn/t}]^\sigma$. The dimension of $\mathbb{Q}\langle G\rangle$ over its center is

$$[\mathbb{Q}\mathcal{A} : \mathbb{Q}[\zeta_{mn/t}]] [\mathbb{Q}[\zeta_{mn/t}] : \mathbb{Q}[\zeta_{mn/t}]^\sigma] = t^2,$$

so the degree of $\mathbb{Q}\langle G\rangle$ is t. □

5.4. REMARK. Suppose x, y is an r-sequence for a 1–complement G with invariant $(m, n, \langle r\rangle)$. We will frequently use the well-known fact that the center of G is $\langle x^t\rangle$ where $t = |r|$, and the commutator subgroup of G is $\langle y\rangle$. The first fact follows easily from the definition of t; the second was proved (briefly) above.

In Proposition 11.1 below we look at the problem of counting the number of isomorphism classes of 1–complements.

CHAPTER 6

Frobenius complements with core index 4

6.1. DEFINITION. Let G be a Frobenius complement with core N. Then there is a unique Frobenius triple Δ and group J from the list (1) of Chapter 1 such that Δ is the invariant of N and $G/N \cong J$ (cf. Theorems 1.4 and 5.2A). We call Δ the *core invariant* of G and J the *type* of G, and say that G is a *J–complement*.

The core index is of course just the order of the type, and hence is one of the integers 1, 4, 12, 24, 60, or 120.

6.2. NOTATION. In this and the next four chapters we will let $\Delta = (m, n, T)$ be a fixed proper Frobenius triple. We will also let r denote a fixed generator of T and set $t = |T|$. In these chapters we will analyze Frobenius complements with core invariant Δ and core index larger than 1.

The reader may find it helpful (but not necessary) to know that in Chapters 6 through 10 p will only be used to denote a rational prime; that a, b, c, d will be elements of $\mathbb{Z}_m$; and that e, f, g, h will often be elements of $\mathbb{Z}_n$ and will always be elements of rings of the form $\mathbb{Z}_s$ where s is a divisor of n.

In this chapter we will analyze all V_4–complements with core invariant $\Delta = (m, n, \langle r \rangle)$. We begin by introducing the numerical objects which will be shown to be in one–to–one correspondence with the isomorphism classes of V_4–complements with core invariant Δ.

6.3. NOTATION. Let $\psi : \mathbb{Z}_m^\bullet \times \mathbb{Z}_{n//t}^\bullet \longrightarrow \mathbb{Z}_{2_{n//t}}^\bullet$ be the natural map (so $\psi(a, f) = f + 2_{n//t}\mathbb{Z}$). Let $\mathcal{S}$ denote the set of all subgroups of $\mathbb{Z}_m^\bullet \times \mathbb{Z}_{n//t}^\bullet$ whose image under ψ is $\langle -1 + 2_{n//t}\mathbb{Z}\rangle$ and which have order at most 4 and exponent at most 2.

The size of $\mathcal{S}$ will be computed in Corollary 6.6 below. Note that $2_{n//t} = 2_n$ if t is odd, which we will find in Theorem 6.5 below to be the only case of interest.

6.4. DEFINITION. Let u, v, x, y be a sequence of elements of a V_4–complement G with core invariant Δ. We will call u, v, x, y an r–sequence for G if x, y is an r–sequence for the core of G (cf. Definition 5.1D); $\langle u, v\rangle$ is a Sylow 2–subgroup of G with $v^2 = x^{n//2}$; and $\langle u, v\rangle$ is contained in the normalizer of $\langle x\rangle$. The *invariant* of an r–sequence u, v, x, y of G is the 4-tuple (a, g, c, h) in $\mathbb{Z}_m^\bullet \times \mathbb{Z}_n^\bullet \times \mathbb{Z}_m^\bullet \times \mathbb{Z}_n^\bullet$ with $uyu^{-1} = y^a$, $uxu^{-1} = x^g$, $vyv^{-1} = y^c$, $vxv^{-1} = x^h$; we call the subgroup $\langle (a, g + n//t\mathbb{Z}), (c, h + n//t\mathbb{Z})\rangle$ of $\mathbb{Z}_m^\bullet \times \mathbb{Z}_{n//t}^\bullet$ a *reduced invariant* of G.

In the next theorem we assert that a V_4–complement has one and only one reduced invariant; neither of these facts is supposed to be obvious at this point. Please note that the choice of r is fixed (cf. Notation 6.2); the uniqueness assertions of Theorem 6.5 below assume this fixed choice of r.

6.5. THEOREM. *There exists a V_4-complement with core invariant Δ if and only if n is even and t is odd. If n is even and t is odd, then every V_4-complement with core invariant Δ has a unique reduced invariant, and assigning to each such Frobenius complement its reduced invariant induces a bijection from the set of isomorphism classes of V_4-complements with core invariant Δ to the set S.*

We will soon give a sequence of lemmas which will contain the proof of Theorem 6.5 and go on to analyze the truncated groups rings. First, however, we use Theorem 6.5 to count isomorphism classes.

6.6. COROLLARY. *Let s denote the number of distinct odd prime divisors of $mn//t$. The number of isomorphism classes of V_4-complements with core invariant Δ is 0 if n is odd or t is even; $2^{s-1}(2^s+1)$ if t is odd and $n \equiv 0 \pmod{4}$; and $(2^s+1)(2^{s-1}+1)/3$ if t is odd and $n \equiv 2 \pmod{4}$.*

PROOF. We may assume without loss of generality that n is even and t is odd (Theorem 6.5), and that $s \geq 1$ (if $s = 0$ then ψ is an isomorphism and S has only one element). The integer $k := mn//2t$ factors into the product of prime powers $\prod_{p|k} p_k$. Each of the groups $\mathbb{Z}_{p_k}^\bullet$ is cyclic [**NZ**, Theorem 2.34, p. 79] and hence has a unique element of order 2. The map ψ of Notation 6.3 is the composition of the natural isomorphism

$$\mathbb{Z}_m^\bullet \times \mathbb{Z}_{n//t}^\bullet \longrightarrow \left(\prod_{p|k} \mathbb{Z}_{p_k}^\bullet\right) \times \mathbb{Z}_{2_n}^\bullet$$

with the projection

$$\left(\prod_{p|k} \mathbb{Z}_{p_k}^\bullet\right) \times \mathbb{Z}_{2_n}^\bullet \longrightarrow \mathbb{Z}_{2_n}^\bullet.$$

Thus S is naturally bijective with the set of subgroups of $\mathbb{Z}^{\bullet s} \times \mathbb{Z}_{2_n}^\bullet$ which project onto the subgroup $\langle -1 + 2_n \mathbb{Z}\rangle$ of $\mathbb{Z}_{2_n}^\bullet$ and have order at most 4 and exponent at most 2.

First suppose $n \equiv 2 \pmod{4}$, so $2_n = 2$ and $\mathbb{Z}_{2_n}^\bullet$ is trivial. Then S is bijective with the set of subgroups of $\mathbb{Z}^{\bullet s}$ of order at most 4. There are 2^s such subgroups of order at most 2, generated by the 2^s elements of $\mathbb{Z}^{\bullet s}$. We have a 6:1 covering of the set of subgroups of order 4 of $\mathbb{Z}^{\bullet s}$ by the set of ordered pairs of distinct nontrivial elements of $\mathbb{Z}^{\bullet s}$ (assign to each such ordered pair the group that its components generate), so there are exactly $(2^s - 1)(2^s - 2)/6$ such subgroups. Thus when $n \equiv 2 \pmod{4}$, then S has cardinality exactly

$$(2^s - 1)(2^s - 2)/6 + 2^s = (2^s + 1)(2^{s-1} + 1)/3.$$

Finally suppose $n \equiv 0 \pmod{4}$, so $-1 + 2_n\mathbb{Z} \neq 1 + 2_n\mathbb{Z}$. Then no group in S is trivial. The groups of order 2 in S correspond bijectively with the elements of $\mathbb{Z}^{\bullet s}$, of which there are 2^s. The groups of order 4 in S correspond bijectively with subgroups of order 4 and exponent 2 of $\mathbb{Z}^{\bullet s} \times \mathbb{Z}_{2_n}^\bullet$ projecting onto $-1 + 2_n\mathbb{Z}$ and each of these is uniquely generated by an unordered pair of distinct nontrivial elements of $\mathbb{Z}^{\bullet s} \times \mathbb{Z}_{2_n}^\bullet$ with last coordinate $-1 + 2_n\mathbb{Z}$. There are exactly $2^s(2^s - 1)/2$ such pairs, and hence S has cardinality

$$2^{s-1}(2^s - 1) + 2^s = 2^{s-1}(2^s + 1).$$

□

We begin the proof of Theorem 6.5 with some lemmas which will be used in this and in later chapters; the first two will be used repeatedly.

6.7. LEMMA. *Let x, y be an r-sequence for a 1-complement with invariant Δ. Suppose $a \in \mathbb{Z}_m$, $g \in \mathbb{Z}_n$, $s \in \mathbb{Z}$ and $(g,t) = 1$. Then $\left(y^a x^g\right)^s = x^{gs}$ if $s \equiv 0$ (mod t), and $\left(y^a x^g\right)^s = y^a x^{gs}$ if $s \equiv 1$ (mod t).*

PROOF. Suppose $s \equiv 0$ (mod t). Then
$$\left(r^g - 1\right)\left(1 + r^g + \cdots + r^{g(s-1)}\right) = r^{sg} - 1 = 0.$$
Since $(g,t) = 1$, then $\langle r \rangle = \langle r^g \rangle$, so $r^g - 1 \in \mathbb{Z}_m^\bullet$ (cf. Lemma 5.3A). Thus $1 + r^g + \cdots + r^{g(s-1)} = 0$. Repeatedly applying the rule $x^g y x^{-g} = y^{r^g}$ we have
$$\left(y^a x^g\right)^s = y^{a(1 + r^g + r^{2g} + \cdots + r^{(s-1)g})} x^{gs} = x^{gs}.$$
Hence if $s \equiv 1$ (mod t) we also have
$$\left(y^a x^g\right)^s = y^a x^g \left(y^a x^g\right)^{s-1} = y^a x^g x^{g(s-1)} = y^a x^{gs}.$$
□

6.8. LEMMA. *Let x, y be an r-sequence for a 1-complement H with invariant Δ. Suppose that $\sigma \in \operatorname{Aut} H$ and that s is a positive integer. Then there exists $a \in \mathbb{Z}_m^\bullet$, $b \in \mathbb{Z}_m$, and $g \in \mathbb{Z}_n^\bullet$ with $\sigma(y) = y^a$, $\sigma(x) = y^b x^g$ and $g \equiv 1$ (mod t). Moreover, if $\sigma^s(x) = x$, then $g^s = 1$ and $b\left(1 + a + \cdots + a^{s-1}\right) = 0$. Finally, if σ has order dividing s and $(s,m) = 1$, then $a^s = 1$ and also $(m, a - 1)$ and $\left(m, 1 + a + \cdots + a^{s-1}\right)$ are relatively prime and have product m.*

PROOF. The existence of a, b and g follows from the definition of an r-sequence and the fact that $\langle y \rangle$ is a characteristic subgroup of H; that a and g are in $\mathbb{Z}_m^\bullet$ and $\mathbb{Z}_n^\bullet$ follows from the fact that σ preserves the orders of elements of H. After all, if some prime p divided both g and n, then by Definition 5.1(A3) and the previous lemma,
$$1 \neq \sigma(x)^{n/p} = \left(y^b x^g\right)^{n/p} = \left(x^n\right)^{g/p} = 1,$$
a contradiction. Thus $g \in \mathbb{Z}_n^\bullet$ and, similarly, $a \in \mathbb{Z}_m^\bullet$.

Since $xy = y^r x$ we have
$$y^{ar} y^b x^g = \sigma(y^r x) = \sigma(xy) = y^b x^g y^a = y^b y^{ar^g} x^g,$$
so $ar + b = b + ar^g$, whence $r^{g-1} = 1$. Then $g \equiv 1$ (mod t). An easy induction argument using the previous lemma shows that for any positive integer j
$$\sigma^j(x) = y^{b(1 + a + \cdots + a^{j-1})} x^{g^j}.$$
Therefore if $\sigma^s(x) = x$, then $b\left(1 + a + \cdots + a^{s-1}\right) = 0$ and $g^s = 1$.

Suppose that $(s,m) = 1$ and σ has order dividing s. Then $y = \sigma^s(y) = y^{a^s}$, so $a^s = 1$. If p is a prime divisor of $(m, a-1)$, then $1 + a + \cdots + a^{s-1} \equiv s$ (mod p), so by hypothesis p does not divide $\left(m, 1 + a + \cdots + a^{s-1}\right)$. Hence $(m, a-1)$ and $\left(m, 1 + a + \cdots + a^{s-1}\right)$ are relatively prime. However since $a^s = 1$ and
$$a^s - 1 = (a - 1)\left(1 + a + \cdots + a^{s-1}\right),$$
then m is the product of $(m, a-1)$ and $\left(m, 1 + a + \cdots + a^{s-1}\right)$. □

6.9. LEMMA. *Suppose $g \in \mathbb{Z}_n$ and $g \equiv 1$ (mod k) for some divisor k of n. If $g^s = 1$ for some integer s relatively prime to k, then $g \equiv 1$ (mod k_n).*

PROOF. Suppose a prime p divides k and p^i divides n. It suffices to show that p^i divides $g-1$, i.e., that $g \equiv 1 \pmod{p^i}$. Since $g \equiv 1 \pmod{p}$, then $1+g+\cdots+g^{s-1}$ is not divisible by p since it is congruent to s modulo p and $(s,k) = 1$. However p^i divides

$$(g-1)(1+g+\cdots+g^{s-1}) = g^s - 1,$$

so p^i divides $g-1$, as required. □

6.10. LEMMA. *If (m, n, T) is the core invariant of a Frobenius complement G which is not a $\mathbb{Z}$-group, then n is even.*

PROOF. Since all the nontrivial groups of display (1) in Chapter 1 have a subgroup isomorphic to V_4, then G has a subgroup which is a V_4-complement with the same core as G. Hence we may suppose without loss of generality that G is a V_4-complement. If n is odd then each Sylow 2-subgroup of G has 4 elements and hence is cyclic [**Pa**, Theorem 18.1(iv), p. 194]. This says G is a $\mathbb{Z}$-group, contradicting our hypothesis. Thus n must be even. □

6.11. LEMMA. *Suppose u, v, x, y are elements of a V_4-complement G with core invariant Δ such that x, y is an r-sequence for the core of G and $\langle u, v \rangle$ is a Sylow 2-subgroup of G with $v^2 = x^{n//2}$. Further suppose a, b, c, $d \in \mathbb{Z}_m$ and g, $h \in \mathbb{Z}_n$ and $uyu^{-1} = y^a$, $uxu^{-1} = y^b x^g$, $vyv^{-1} = y^c$ and $vxv^{-1} = y^d x^h$. Then t is odd, $a^2 = c^2 = 1$, $g^2 = h^2 = 1$, $g \equiv h \equiv 1 \pmod{t}$, $b(a+1) = d(c+1) = 0$, $d(a-1) = b(c-1)$, and $h \equiv -g \equiv 1 \pmod{2_n}$.*

PROOF. Since m is odd (Lemma 5.3(B1)) and n is even (Lemma 6.10) and G is a V_4-complement, then $\langle u, v \rangle$ is a generalized quaternion group of order $4(2_n)$ and v has order $2(2_n)$. Hence u has order 4 (Lemma 4.1A). Hence conjugation by u induces an automorphism of $\langle x, y \rangle$ of order dividing 2. Thus by Lemma 6.8, $a^2 = 1$, $g^2 = 1$, $g \equiv 1 \pmod{t}$, and $b(a+1) = 0$. The lemma (applied to conjugation by v) also implies that $h \equiv 1 \pmod{t}$. Since $v^2 = x^{n//2}$,

$$y^{c^2} = v^2 y v^{-2} = x^{n//2} y x^{-n//2} = y^{r^{n//2}},$$

so $c^2 = r^{n//2}$. Also $x^{n//2} uv = v^2 v^{-1} u = vu$, so

$$y^{ac} = vuy(vu)^{-1} = y^{car^{n//2}},$$

so $ac = acr^{n//2}$. But then $r^{n//2} = 1$, so $c^2 = 1$ and t divides $n//2$. Thus t is odd and so conjugation by v has order dividing 2 on $\langle x, y \rangle$. Hence $h^2 = 1$ and $d(c+1) = 0$. Next since $uv = v^{-2}vu$ and $v^2 = x^{n//2}$ commutes with x and y, we have

$$y^{da+b} x^{hg} = uvxv^{-1}u^{-1} = vuxu^{-1}v^{-1} = y^{bc+d} x^{hg},$$

so $da + b = bc + d$, i.e., $d(a-1) = b(c-1)$. Since $v^2 = x^{n//2}$, then by Lemma 6.7,

$$x^{n//2} = vx^{n//2}v^{-1} = \left(y^d x^h\right)^{n//2} = x^{h(n//2)}.$$

Thus $n//2 \equiv h(n//2) \pmod{n}$, so $h \equiv 1 \pmod{2_n}$. Similarly,

$$x^{g(n/2)} = ux^{n//2}u^{-1} = uv^2 u^{-1} = v^{-2} = x^{-n//2},$$

so $g \equiv -1 \pmod{2_n}$. □

6.12. REMARK AND NOTATION. Suppose that u, v, x, y is an r-sequence for a V_4-complement G with core invariant Δ. Then t is odd by Lemma 6.11. Let

$$\theta = \theta_{u,v,x,y} : \langle u, v \rangle \longrightarrow \mathbb{Z}_m^\bullet \times \mathbb{Z}_{n//t}^\bullet$$

assign to each $w \in \langle u, v \rangle$ the pair $(a, g + n//t\mathbb{Z})$ where $a \in \mathbb{Z}_m$ and $g \in \mathbb{Z}_n$ satisfy $wyw^{-1} = y^a$ and $wxw^{-1} = x^g$. Note that θ is a homomorphism, $v^2 = x^{n//2} \in \ker \theta$, and the image of θ is exactly the reduced invariant of G corresponding to u, v, x, y. In particular θ and the reduced invariant depend only on the group $\langle u, v \rangle$ and not on the sequence u, v.

We are now ready to prove about half of Theorem 6.5.

6.13. LEMMA. *If there is a V_4-complement with core invariant Δ, then n is even and t is odd. Every V_4-complement with core invariant Δ has a unique reduced invariant and it is in $\mathcal{S}$. Moreover, two V_4-complements with core invariant Δ are isomorphic if and only if they have the same reduced invariant.*

Note that to say a V_4-complement with core invariant Δ has a reduced invariant is to say it has an r-sequence; that the reduced invariant is unique says that all the r-sequences give rise to the same reduced invariant.

PROOF. Suppose G is a V_4-complement with core invariant Δ. Then n is even by Lemma 6.10. There exists an r-sequence x, y for the core of G (Theorem 5.2B). There is a Sylow 2-subgroup of G containing $x^{n//2}$; it is a generalized quaternion group $\langle u, v \rangle$ of order $(2_n)4$ with $u^2 = v^{2_n}$, $uvu^{-1} = v^{-1}$, and $u^4 = 1$. If $2_n = 2$, then $x^{n//2} = v^2$ (G has a unique element of order 2 [**Pa**, Theorem 18.1(iii), p. 194]). Suppose $2_n > 2$. Then

$$\langle x^{n//2} \rangle = \langle x, y \rangle \cap \langle u, v \rangle \triangleleft \langle u, v \rangle,$$

so by Lemma 4.1C (applied with $2^s = 2_n$), $\langle x^{n//2} \rangle = \langle v^2 \rangle$, so for some odd integer k, $x^{n//2} = (v^k)^2$. Replacing v by v^k we may assume without loss of generality that v is chosen with $v^2 = x^{n//2}$. Since $\langle x, y \rangle$ and $\langle y \rangle$ are normal subgroups of G, there exist a, b, c, $d \in \mathbb{Z}_m$ and g, $h \in \mathbb{Z}_n$ with

$$uyu^{-1} = y^a, \quad uxu^{-1} = y^b x^g, \quad vyv^{-1} = y^c, \quad vxv^{-1} = y^d x^h.$$

Then by Lemma 6.11, t is odd and $a^2 = c^2 = 1$, $g^2 = h^2 = 1$, $g \equiv h \equiv 1 \pmod{t}$, $b(a+1) = d(c+1) = 0$, $d(a-1) = b(c-1)$ and $h \equiv -g \equiv 1 \pmod{2_n}$. Conjugation by u and by v each induce on $\langle x, y \rangle$ automorphisms of order dividing 2 (since t is odd, $v^2 = x^{n//2}$ is a power of x^t, which commutes with y since $r^t = 1$). Hence by Lemma 6.8

$$m = (m, a+1)(m, a-1) = (m, c+1)(m, c-1)$$

where the two factors of m in both cases are relatively prime. Hence there exist a', $c' \in \mathbb{Z}_m^\bullet$ with

$$a'(a-1) \equiv 1 \pmod{(m, a+1)} \text{ and } c'(c-1) \equiv 1 \pmod{(m, c+1)}.$$

Since $d(a-1) = b(c-1)$ we have

$$a'b \equiv c'd \pmod{(m, c+1, a+1)}.$$

Thus there exists an integer j with

$$j \equiv -a'b \pmod{(m, a+1)} \text{ and } j \equiv -c'd \pmod{(m, c+1)},$$

whence
$$(a-1)j + b \equiv 0 \pmod{(m, a+1)}$$
and
$$(c-1)j + d \equiv 0 \pmod{(m, c+1)}.$$
But then
$$0 \equiv (a-1)j + b \equiv (c-1)j + d \pmod{m}$$
since the equations $b(a+1) = d(c+1) = 0$ imply
$$b \equiv 0 \pmod{(m, a-1)} \text{ and } d \equiv 0 \pmod{(m, c-1)}.$$

The sequence u, v, $y^j x$, y is an r–sequence for G with invariant (a, g, c, h) because $(y^j x)^n = 1$ and $(y^j x)y(y^j x)^{-1} = y^r$; $(y^j x)^{n//2} = x^{n//2} = v^2$ (Lemma 6.7); $uy^j x u^{-1} = y^{ja+b} x^g = (y^j x)^g$ since $ja + b \equiv j \pmod{m}$; and $vy^j x v^{-1} = y^{jc+d} x^h = (y^j x)^h$ since $jc + d \equiv j \pmod{m}$. The corresponding reduced invariant is $\langle (a, g + (n//t)\mathbb{Z}), (c, h + (n//t)\mathbb{Z}) \rangle$, which is in $\mathcal{S}$ by the identities proved earlier. Thus we have shown that any V_4-complement with core invariant Δ has an r-sequence, and its corresponding reduced invariant is in $\mathcal{S}$.

We need some notation. Suppose G and G_1 are V_4-complements with core invariant Δ. Suppose u, v, x, y and u_1, v_1, x_1, y_1 are r-sequences with invariants (a, g, c, h) and (a_1, g_1, c_1, h_1) and corresponding reduced invariants S and S_1 for G and G_1, respectively.

We now prove G has a unique reduced invariant. For this we use the above notation, taking $G = G_1$. There exists $z \in G$ with $z \langle u_1, v_1 \rangle z^{-1} = \langle u, v \rangle$. We have integers ω, δ, μ, ρ, σ, τ with $x_1 = y^\omega x^\delta$, $y_1 = y^\mu$, $zyz^{-1} = y^\rho$, and $zxz^{-1} = y^\sigma x^\tau$. By hypothesis, $y_1^r = x_1 y_1 x_1^{-1} = y_1^{r^\delta}$, so $r = r^\delta$. Thus $\delta \equiv 1 \pmod{t}$. Also $\tau \equiv 1 \pmod{t}$ by Lemma 6.8. Hence by Lemma 6.7
$$zx_1 z^{-1} = y^{\rho\omega}(y^\sigma x^\tau)^\delta = y^{\rho\omega + \sigma} x^{\tau\delta}.$$
Note that $zu_1 z^{-1}$, $zv_1 z^{-1}$, $zx_1 z^{-1}$, $zy_1 z^{-1}$ must be an r-sequence for G with the same invariant as u_1, v_1, x_1, y_1. Next calculate that
$$(zx_1 z^{-1})^{n//2} = (y^{\rho\omega+\sigma} x^{\tau\delta})^{n//2} = x^{\tau\delta(n//2)} = (v^{\delta\tau})^2.$$
But τ and δ are odd, since they are relatively prime to n. Thus
$$\langle zu_1 z^{-1}, zv_1 z^{-1} \rangle = \langle u, v \rangle = \langle u, v^{\delta\tau} \rangle.$$
Hence u, $v^{\delta\tau}$, $zx_1 z^{-1}$, $zy_1 z^{-1}$ is also an r-sequence for G with possibly a different invariant than u_1, v_1, x_1, y_1 but with the same reduced invariant S_1 (cf. Remark 6.12). We now compute the invariant for this r-sequence. First note that $uzy_1 z^{-1} u^{-1} = y^{\mu\rho a} = (zy_1 z^{-1})^a$; similarly
$$v^{\delta\tau} zy_1 z^{-1} v^{-\delta\tau} = (zy_1 z^{-1})^{c^{\delta\tau}} = (zy_1 z^{-1})^c.$$
(Note that $c^2 = 1$ and $\delta\tau$ is odd, so $c^{\delta\tau} = c$.) Next, modulo the normal subgroup $\langle y \rangle$ we have
$$uzx_1 z^{-1} u^{-1} \equiv x^{\delta\tau g} \equiv (zx_1 z^{-1})^g$$
and similarly
$$v^{\delta\tau} zx_1 z^{-1} v^{-\delta\tau} \equiv vzx_1 z^{-1} v^{-1} \equiv (zx_1 z^{-1})^h.$$

Since $v^{\delta\tau}$ and u are in the normalizer of $\langle zx_1z^{-1}\rangle$ we therefore have the equations
$$uzx_1z^{-1}u^{-1} = (zx_1z^{-1})^g \quad \text{and} \quad v^{\delta\tau}(zx_1z^{-1})v^{-\delta\tau} = (zx_1z^{-1})^h.$$
Thus u, $v^{\delta\tau}$, zx_1z^{-1}, zy_1z^{-1} has the same invariant (a,g,c,h) as u, v, x, y, and hence the same reduced invariant S. Thus $S = S_1$, as claimed.

Let us now assume that $S = S_1$. We complete the proof of the lemma by showing that $G \cong G_1$. (We use the notation introduced in the paragraph before the last one.) It suffices to show G and G_1 have r–sequences with the same invariant, since then they have equivalent presentations and hence are isomorphic. First suppose $2_n > 2$, so $1 + 2_n\mathbb{Z} \neq -1 + 2_n\mathbb{Z}$ and $|S|$ is 2 or 4. Let $\theta = \theta_{u,v,x,y}$ and $\theta_1 = \theta_{u_1,v_1,x_1,y_1}$ (cf. Remark 6.12). If $|S| = 2$, then by Lemma 6.11 $\theta(u)$ must be the unique nontrivial element of S and $\theta(v)$ must be trivial; the same is true for $\theta_1(u_1)$ and $\theta_1(v_1)$, so $\theta(u) = \theta_1(u_1)$ and $\theta(v) = \theta_1(v_1)$. On the other hand if $|S| = 4$, then $\theta(v) = \theta_1(v_1)$ since both must be the unique nontrivial element of $S \cap \ker\psi$ (cf. Lemma 6.11 and Remark 6.12; ψ is defined in Notation 6.3). Also $\theta(u)$ and $\theta_1(u_1)$ are not in $\ker\psi$ so either $\theta(u) = \theta_1(u_1)$ or $\theta(u) = \theta_1(u_1v_1)$; we may assume without loss of generality that $\theta(u) = \theta_1(u_1)$ (if necessary replace u_1 by u_1v_1 which changes the invariant but not the reduced invariant of the r–sequence u_1, v_1, x_1, y_1). Hence whatever the value of $|S|$, we have $a = a_1$, $c = c_1$, $g \equiv g_1 \pmod{n//t}$ and $h \equiv h_1 \pmod{n//t}$. But by Lemma 6.9 (applied with $k = t$), $g \equiv g_1 \equiv h \equiv h_1 \pmod{t_n}$. Thus $g = g_1$ and $h = h_1$, so G and G_1 have r–sequences with the same invariant, as was to be proved.

Next suppose $2_n = 2$. If w and z are any pair of generators of $\langle u,v\rangle$, then w, z, x, y is an r–sequence for G with reduced invariant S. With such a modification we may assume without loss of generality that u and v are chosen with $\theta(u) = \theta_1(u_1)$ and $\theta(v) = \theta_1(v_1)$. Then $a = a_1$, $g \equiv g_1 \pmod{n//t}$, $c = c_1$, and $h \equiv h_1 \pmod{n//t}$. Again applying Lemma 6.9 we have $(a,g,c,h) = (a_1,g_1,c_1,h_1)$, which completes the proof that $G \cong G_1$. □

In the next theorem we complete the proof of Theorem 6.5 by showing that when n is even and t odd, every element of $\mathcal{S}$ is the reduced invariant of some V_4–complement with core invariant Δ. As with earlier constructions of Frobenius complements we will begin by constructing the truncated group ring. The reader might wish to read the last paragraph of the next theorem before reading the rest.

6.14. THEOREM. *Suppose n is even and t is odd. Let $S \in \mathcal{S}$. Then there exists* $(a,g,c,h) \in \mathbb{Z}_m^\bullet \times \mathbb{Z}_n^\bullet \times \mathbb{Z}_m^\bullet \times \mathbb{Z}_n^\bullet$ *with*
$$S = \langle (a, g + n//t\mathbb{Z}), (c, h + n//t\mathbb{Z})\rangle;$$
$g \equiv h \equiv 1 \pmod{t_n}$; $h \equiv -g \equiv 1 \pmod{2_n}$; *and* $(c,h) = (1,1)$ *if* $|S| < 4$.

Let $\mathcal{A} = (\mathbb{Z}[\zeta_{mn/t}], \sigma, t, \zeta_{n/t})$ *be as in Theorem 5.2C, and set* $x = \hat{\sigma}$ *and* $y = \zeta_m$. *Then there exists* $\sigma_0 \in \operatorname{Aut}\mathcal{A}$ *with* $\sigma_0(y) = y^c$ *and* $\sigma_0(x) = x^h$. *Further,* $\sigma_0(x^{n//2}) = x^{n//2}$, $|\sigma_0| \leq 2$, *and* $x^{n//2}z = \sigma_0^2(z)x^{n//2}$ *for all* $z \in \mathcal{A}$. *Let* $\mathcal{A}_1 = (\mathcal{A}, \sigma_0, 2, x^{n//2})$ *and* $v = \hat{\sigma}_0$. *Then there exists* $\sigma_1 \in \operatorname{Aut}\mathcal{A}_1$ *with* $|\sigma_1| = 2$, $\sigma_1(y) = y^a$, $\sigma_1(x) = x^g$, *and* $\sigma_1(v) = v^{-1}$. *Let* $\mathcal{B} = (\mathcal{A}_1, \sigma_1, 2, -1)$ *and* $u = \hat{\sigma}_1$.

The multiplicative monoid $G := \langle x, y, u, v\rangle$ *is a V_4–complement with core invariant Δ; u, v, x, y is an r–sequence for G with invariant (a,g,c,h) and reduced invariant S; and the inclusion* $G \longrightarrow \mathcal{B}$ *induces an isomorphism* $\mathbb{Z}\langle G\rangle \longrightarrow \mathcal{B}$. $\mathbb{Z}\langle G\rangle$ *is free of rank $\phi(|G|) = 4\phi(mn)$ as a $\mathbb{Z}$–module.*

The conditions specified on the automorphisms σ_0 and σ_1 should be checked to imply that the rings $\mathcal{A}_1$ and $\mathcal{B}$ above make sense (cf. Theorem 2.5 and Remark 2.6B).

PROOF. The existence of (a, g, c, h) follows from the Chinese Remainder Theorem (note t_n and $n//t$ are relatively prime with product n) and the definition of $\mathcal{S}$. One easily checks that x^h, y^c is an r–sequence for the 1–complement $H := \langle x, y \rangle$ (recall that $h \equiv 1 \pmod{t}$, so $r^h = r$). Hence there exists an automorphism of H mapping x to x^h and y to y^c; this map induces an automorphism $\widetilde{\sigma}_0$ of $\mathbb{Z}\langle H \rangle$ taking $\overline{x}$ to $\overline{x}^h$ and $\overline{y}$ to $\overline{y}^c$ and hence an automorphism σ_0 of $\mathcal{A}$ taking x to x^h and y to y^c (cf. Theorem 5.2C). Note $\sigma_0(x^{n//2}) = x^{h(n//2)} = x^{n//2}$ since $h \equiv 1 \pmod{2_n}$, and $|\sigma_0| \leq 2$ since $h^2 \equiv 1$ and $c^2 \equiv 1$ (cf. the definition of $\mathcal{S}$ and the fact that $h \equiv 1 \pmod{t_n}$). Also since t is odd, $x^{n//2}$ is in the center of $\langle x, y \rangle$, so $x^{n//2}z = \sigma_0^2(z)x^{n//2}$ for all $z \in \mathcal{A}$. Hence it makes sense to define $\mathcal{A}_1 = (\mathcal{A}, \sigma_0, 2, x^{n//2})$ and to let $v = \widehat{\sigma}_0$. Let $H_1 = \langle x, y, v \rangle$. By the Cyclic Extension Lemma 2.8 H_1 is a Frobenius complement and the inclusion $H_1 \longrightarrow \mathcal{A}_1$ induces an isomorphism $\mathbb{Z}\langle H_1 \rangle \longrightarrow \mathcal{A}_1$. (The isomorphism in question is actually the composition of the isomorphism $\mathbb{Z}\langle H_1 \rangle \longrightarrow (\mathbb{Z}\langle H \rangle, \widetilde{\sigma}_0, 2, \overline{x}^{n//2})$ from Lemma 2.8 and an isomorphism

$$\left(\mathbb{Z}\langle H \rangle, \widetilde{\sigma}_0, 2, \overline{x}^{n//2}\right) \longrightarrow (\mathcal{A}, \sigma_0, 2, x^{n//2})$$

induced by the isomorphism $\mathbb{Z}\langle H \rangle \longrightarrow \mathcal{A}$ of Theorem 5.2C.) Since $\mathcal{A}$ is free of rank $\phi(mn)$ as a $\mathbb{Z}$-module, $\mathcal{A}_1$ is free of rank $2\phi(mn)$. H_1 has a presentation with generators x, y, v satisfying the relations

$$x^n = y^m = 1, \quad xy = y^r x, \quad v^2 = x^{n//2}, \quad vx = x^h v \quad \text{and} \quad vy = y^c v.$$

(The definition of $\mathcal{A}_1$ is used here; for example $vxv^{-1} = \widehat{\sigma}_0 x \widehat{\sigma}_0^{-1} = \sigma_0(x)\widehat{\sigma}_0\widehat{\sigma}_0^{-1} = x^h$.) Now H_1 is also generated by x^g, y^a, v^{-1} and these generators can also be checked to satisfy the above relations: $(x^g)^n = (y^a)^m = 1$; $x^g y^a x^{-g} = y^{ar}$ (since $g \equiv 1 \pmod{t}$); $(v^{-1})^2 = x^{-n//2} = (x^g)^{n//2}$ (since $g \equiv -1 \pmod{2_n}$);

$$v^{-1}y^a(v^{-1})^{-1} = v^{-2}vy^a v^{-1}v^2 = x^{-n//2}y^{ac}x^{n//2} = (y^a)^c;$$

and, similarly, $v^{-1}x^g(v^{-1})^{-1} = (x^g)^h$. Hence there exists an automorphism of H_1 taking x, y and v to x^g, y^a and v^{-1}, respectively; this induces an automorphism $\widetilde{\sigma}_1$ of $\mathbb{Z}\langle H_1 \rangle$ and hence induces an automorphism σ_1 of $\mathcal{A}_1$ taking x, y and v to x^g, y^a and v^{-1} respectively (arguing as above in the construction of σ_0). Since $a^2 = 1$, $g^2 = 1$ and $\sigma_1(v) \neq v$, and $\mathcal{A}_1$ is generated as a ring by x, y and v, therefore $|\sigma_1| = 2$. Hence we can form $\mathcal{B} := (\mathcal{A}_1, \sigma_1, 2, -1)$ and set $u = \widehat{\sigma}_1$ and let $G = \langle x, y, u, v \rangle$. $\langle u, v \rangle$ is a Sylow 2-subgroup of G and it is a generalized quaternion group, so it has a unique subgroup of order 2. Then by the Cyclic Extension Lemma 2.8 G is a Frobenius complement and the inclusion $G \longrightarrow \mathcal{B}$ induces an isomorphism $\mathbb{Z}\langle G \rangle \longrightarrow \mathcal{B}$ (as before, Lemma 2.8 gives an isomorphism $\mathbb{Z}\langle G \rangle \longrightarrow (\mathbb{Z}\langle H_1 \rangle, \widetilde{\sigma}_1, 2, -1)$ and the isomorphism $\mathbb{Z}\langle H_1 \rangle \longrightarrow \mathcal{A}_1$ induces an isomorphism $(\mathbb{Z}\langle H_1 \rangle, \widetilde{\sigma}_1, 2, -1) \longrightarrow \mathcal{B}$). Now since $\mathcal{A}_1$ is free of rank $2\phi(mn)$ as a $\mathbb{Z}$-module, then $\mathcal{B}$ is free of rank $4\phi(mn)$ as a $\mathbb{Z}$-module. By construction H is a normal $\mathbb{Z}$-subgroup of G and $G/H = \langle uH, vH \rangle \cong V_4$ since u^2 and v^2 are in H and $[G : H] = 4$. Thus H is the core of G, and G has core index 4 and core invariant Δ. Moreover u, v, x, y is by construction an r-sequence for G with invariant (a, g, c, h) and hence with reduced invariant $\langle(a, g + n//t\mathbb{Z}), (c, h + n//t\mathbb{Z})\rangle = S$. Finally, since n is even and $[G : H] = 4$, then the rank of $\mathbb{Z}\langle G\rangle$ is $4\phi(mn) = \phi(4mn) = \phi(|G|)$. □

The last theorem in this chapter gives some basic properties of the rational truncated group ring of a V_4–complement. We will use the notation of the previous theorem, so that in particular G is a V_4–complement with core invariant Δ and reduced invariant S, and $\mathbb{Z}\langle G\rangle \cong \mathcal{B}$.

6.15. THEOREM. $\mathbb{Q}\langle G\rangle$ is a simple algebra with dimension $\phi(|G|) = 4\phi(mn)$ over $\mathbb{Q}$ and with degree $4t$ if $|S| = 4$ and $2t$ otherwise. The inclusion $G \longrightarrow \mathcal{B}$ induces an isomorphism $\mathbb{Q}\langle G\rangle \longrightarrow \mathbb{Q}\mathcal{B}$, and the center of $\mathbb{Q}\mathcal{B}$ is K^{σ_0,σ_1} if $|S| = 4$ and $K[v]^{\sigma_1}$ if $|S| \neq 4$, where K denotes the center of $\mathbb{Q}\mathcal{A}$.

The proof of Theorem 6.15 will show that if $|S| \neq 4$, then $K[v]$ is a quadratic field extension of K and the restriction of σ_1 to it has order 2. If $|S| = 1$, then $K[v]^{\sigma_1} = K$ and the proof will show that $K[u, v]$ is a quaternion algebra over K and $\mathbb{Q}\mathcal{B} \cong \mathbb{Q}\mathcal{A} \otimes_K K[u, v]$.

PROOF. $[\mathbb{Q}\langle G\rangle : \mathbb{Q}] = 4\phi(mn)$ since $\mathbb{Z}\langle G\rangle$ is free of rank $4\phi(mn)$ as a $\mathbb{Z}$–module. We may assume by Theorem 6.5 that G is the group of Theorem 6.14. First suppose $|S| = 1$. Then $2_n = 2$ (otherwise S must have nontrivial image in $\mathbb{Z}_{2_n}$). Thus $K[u, v]$ is a quaternion algebra over K (note $u^2 = v^2 = -1$ and $uvu^{-1} = v^{-1}$) and hence is simple with center K and dimension 4 over K. By Theorem 5.2D, $\mathbb{Q}\mathcal{A}$ is central simple with dimension t^2 over K. Further, since $|S| = 1$, then u and v commute with x and y, and so every element of $K[u, v]$ commutes with every element of $\mathbb{Q}\mathcal{A}$. Also $\mathbb{Q}\mathcal{B}$ is clearly generated as a $\mathbb{Q}$–algebra by x, y, u and v, and

$$[\mathbb{Q}\mathcal{B} : K] = 4[\mathbb{Q}\mathcal{A} : K] = [K[u, v] : K][\mathbb{Q}\mathcal{A} : K].$$

Hence $\mathbb{Q}\langle G\rangle \cong \mathbb{Q}\mathcal{B} \cong K[u, v] \otimes_K \mathbb{Q}\mathcal{A}$ by [**P**, Proposition c, p. 165]. Thus $\mathbb{Q}\langle G\rangle$ is a central simple K–algebra of degree $2t$ [**P**, Proposition b(i), p. 226]. Since $2_n = 2$, $K[v]$ is clearly a quadratic field extension of K with fixed field K under σ_1 (recall that $\sigma_1(v) = -v$).

Next suppose $|S| = 2$. Then $(c, h) = (1, 1)$, so v commutes with x and y. $L := \mathbb{Q}[\zeta_{mn/t}]$ has no root of unity of order $2(2_n) = |v|$, so v generates a quadratic field extension of L and hence of K (note $v^2 = x^{n//2} \in K$ since t is odd). By [**P**, Proposition c, p. 165], $\mathbb{Q}\mathcal{A}_1 \cong \mathbb{Q}\mathcal{A} \otimes_K K[v]$, so $\mathbb{Q}\mathcal{A}_1$ is simple with center $K[v]$ [**P**, Proposition b(ii), p. 226]. The automorphism σ_1 of $\mathbb{Q}\mathcal{A}_1$ induces on $K[v]$ an automorphism of order 2 (recall that $\sigma_1(v) = v^{-1}$). Hence $\mathbb{Q}\mathcal{B}$ is simple with center $K[v]^{\sigma_1}$ by the theorem of Albert (Theorem 2.5). The degree is therefore $2t$.

Finally suppose $|S| = 4$. Then the surjection $\theta_{u,v,x,y} : \langle u, v\rangle \longrightarrow S$ induces an isomorphism $\theta : \langle u, v\rangle/\langle v^2\rangle \longrightarrow S$ (cf. Remark 6.12). Since v^2 is in the center of $\langle x, y\rangle$, conjugation induces a homomorphism $\delta : \langle u, v\rangle/\langle v^2\rangle \longrightarrow \text{Aut}\langle x^t, y\rangle$. Suppose $w \in \langle u, v\rangle$ commutes with x^t and y, i.e., that $w\langle v^2\rangle \in \ker \delta$. Write $wxw^{-1} = x^f$ for $f \in \mathbb{Z}_n^{\bullet}$ (recall that u, v, x, y is an r–sequence). Then $wyw^{-1} = y$ and $x^t = wx^tw^{-1} = x^{ft}$, so $f \equiv 1 \pmod{n/t}$. Hence $\theta(w\langle v^2\rangle) = (1, f + n//t\mathbb{Z}) = (1, 1)$, so $w\langle v^2\rangle$ is trivial. Therefore δ is injective. Hence the image of δ, which is exactly the group of automorphisms of $\langle y, x^t\rangle = \langle \zeta_m, \zeta_{n/t}\rangle = \langle \zeta_{mn/t}\rangle$ induced by σ_0 and σ_1, has 4 elements. Thus the group $\langle \sigma_0|L, \sigma_1|L\rangle$ of automorphisms of $L := \mathbb{Q}[\zeta_{mn/t}]$ has 4 elements, so $[L : L^{\sigma_0,\sigma_1}] = 4$. Since $[L : K] = t$ is odd, therefore $[L : L^{\sigma_0,\sigma_1} \cap K] = 4t$. Thus σ_0 induces an automorphism of order 2 on K and σ_1

induces an automorphism of order 2 on K^{σ_0}. Therefore $\mathbb{Q}\mathcal{A}_1$ is simple with center K^{σ_0} and $\mathbb{Q}\mathcal{B}$ is simple with center K^{σ_0,σ_1} (cf. Theorem 5.2D and Theorem 2.5). The degree of $\mathbb{Q}\mathcal{B}$ over its center is therefore $4t$. □

CHAPTER 7

Frobenius complements with core index 12

The isomorphism classes of A_4–complements and their truncated group rings are computed here. We begin by introducing the numerical objects which will be shown to be bijective with isomorphism classes of A_4–complements with core invariant Δ. We continue to use the notation of Notation 6.2.

7.1. NOTATION. Let $\mathcal{S}$ denote the set of all cyclic subgroups S of $\mathbb{Z}_m^\bullet \times \mathbb{Z}_{n//6t}^\bullet$ such that $T^* \times \langle 1 \rangle$ is a subgroup of S of index either $(3,t)$ or $(3,n)$, where T^* denotes the Sylow 3–subgroup of T.

In some cases $\mathcal{S}$ can be described much more simply; see Remark 7.4 below. Notice that the meaning of the symbol $\mathcal{S}$ here is different from that in Chapter 6.

7.2. DEFINITION. Let G be an A_4–complement with core invariant Δ. Suppose that $(d, h) \in \mathbb{Z}_m^\bullet \times \mathbb{Z}_n^\bullet$ and that u, v, x, y, z is a sequence of elements of G such that $H := \langle u, v, x, y \rangle$ is a V_4–complement with the same core as G; u, v, x, y is an r–sequence for H with trivial invariant; $z^3 = x^{n//3}$, $zyz^{-1} = y^d$, $zxz^{-1} = x^h$, $zvz^{-1} = u$, and $zuz^{-1} = vu$. We then call u, v, x, y, z an r–sequence for G with invariant (d, h) and reduced invariant $\langle (d, h + n//6t\mathbb{Z}) \rangle$ (so a reduced invariant of G is a subgroup of $\mathbb{Z}_m^\bullet \times \mathbb{Z}_{n//6t}^\bullet$).

The next theorem says (among other things) that an A_4–complement with core invariant Δ has an r–sequence and hence a reduced invariant, and that all r–sequences give rise to the same reduced invariant. Note that the notions of r–sequences for 1–complements, V_4–complements, and A_4–complements are three distinct (but related) concepts.

7.3. THEOREM. *If there exists an A_4–complement with core invariant Δ, then $n \equiv 2 \pmod 4$. If $n \equiv 2 \pmod 4$, then every A_4–complement with core invariant Δ has a unique reduced invariant and assigning to each such Frobenius complement its reduced invariant induces a bijection from the set of isomorphism classes of A_4–complements with core invariant Δ to the set $\mathcal{S}$.*

The reader may find it useful to keep in mind that

$$n \equiv 2 \pmod 4 \iff (n, 4) = 2 \iff 2_n = 2 \iff n//2 = n/2\,.$$

Before turning to the proof of Theorem 7.3 we give an analysis of the set $\mathcal{S}$ in the next remark, and then use this remark and Theorem 7.3 to count the number of isomorphism classes of A_4–complements with core invariant Δ. The computation will show that $\mathcal{S}$ can be empty even if $n \equiv 2 \pmod 4$, so this condition is not sufficient for the existence of an A_4–complement with with core invariant Δ.

7.4. REMARK. If 3 does not divide n, then $\mathcal{S}$ contains only the trivial group. If 3 divides n but not t, then $\mathcal{S}$ consists of all subgroups of $\mathbb{Z}_m^\bullet \times \mathbb{Z}_{n//6t}^\bullet$ of order dividing

3. Finally, if 3 divides t, then $\mathcal{S}$ consists of all cyclic subgroups of $\mathbb{Z}_m^\bullet \times \mathbb{Z}_{n//6t}^\bullet$ having $\langle (r^{n//3}, 1) \rangle$ as a subgroup of index 3. In this case $\mathcal{S}$ is naturally bijective with the set of cosets of the factor group $(\mathbb{Z}_m^\bullet \times \mathbb{Z}_{n//6t}^\bullet)/\langle (r^{t/3}, 1) \rangle$ of the form $(a,g)\langle (r^{t/3}, 1) \rangle$ where $a^3 = r^{n//3}$ and $g^3 = 1$. (Assign to any such coset the subgroup of $\mathbb{Z}_m^\bullet \times \mathbb{Z}_{n//6t}^\bullet$ generated by any of its three members.)

PROOF. The assertions above are easily verified except for the last one. Suppose 3 divides t. Any member of $\mathcal{S}$ is generated by an element (a,g) of $\mathbb{Z}_m^\bullet \times \mathbb{Z}_{n//6t}^\bullet$ such that $g^3 = 1$ and $a^3 = r^{n//3}$ (if (a,g) is any generator, then $(a,g)^3$ generates $\langle (r^{n//3}, 1) \rangle$, so for some $s \in \mathbb{Z}$ not divisible by 3 we have $(a,g)^{3s} = (r^{n//3}, 1)$; now just replace (a,g) by (a^s, g^s)). For any $s \in \mathbb{Z}$ the cyclic group generated by $(a,g)(r^{t/3}, 1)^s$ is independent of the value of s since $(r^{t/3}, 1) \in \langle (r^{n//3}, 1) \rangle \subset \langle (a,g) \rangle$. If $\langle (b,h) \rangle = \langle (a,g) \rangle$ and $h^3 = 1$ and $b^3 = r^{n//3}$, then $((a,g)(b,h)^{-1})^3 = 1$, so for some integer s, $(a,g) = (b,h)(r^{t/3}, 1)^s$ (note that $\langle (r^{t/3}, 1) \rangle$ is the unique subgroup of $\langle (a,g) \rangle$ of order 3). $\square$

7.5. COROLLARY. *Let s denote the number of prime factors p of $mn//t$ with $p \equiv 1 \pmod{3}$. The number of isomorphism classes of A_4-complements with core invariant Δ is 1 if $n \equiv 2 \pmod{4}$ and 3 does not divide n; $(3^s + 1)/2$ if $n \equiv 2 \pmod{4}$ and 3 divides n but not t; 3^{s-1} if $n \equiv 2 \pmod{4}$, 3 divides t, and*

(4) $\qquad$ *if $p \mid m$ (p prime) and $3_{p-1} = |r^{n//3} + p\mathbb{Z}|$, then $3_{p-1} = 1$;*

and 0 otherwise (i.e., if either $n \not\equiv 2 \pmod{4}$ or else 3 divides t and for some prime divisor p of m, $1 < 3_{p-1} = |r^{n//3} + p\mathbb{Z}|$).

Note that the condition $1 < 3_{p-1} = |r^{n//3} + p\mathbb{Z}|$ simply says that $\langle r + p\mathbb{Z} \rangle$ contains a nontrivial Sylow 3-subgroup of $\mathbb{Z}_p^\bullet$.

PROOF. We may assume without loss of generality that $n \equiv 2 \pmod{4}$ and that 3 divides n, since otherwise the Corollary is immediate from Theorem 7.3. For any finite abelian group A and power 3^i let A_{3^i} denote the set of elements of A of order dividing 3^i. By the Chinese Remainder Theorem, from the prime factorization

$$mn//6t = \prod_p p^{i(p)}$$

(product over all prime factors p of $mn//6t$) we obtain a ring isomorphism

$$\mathbb{Z}_m \times \mathbb{Z}_{n//6t} \longrightarrow \prod_p \mathbb{Z}_{p^{i(p)}}$$

and hence a group isomorphism

$$\mathbb{Z}_m^\bullet \times \mathbb{Z}_{n//6t}^\bullet \longrightarrow \prod_p \mathbb{Z}_{p^{i(p)}}^\bullet .$$

Each factor $\mathbb{Z}_{p^{i(p)}}^\bullet$ is cyclic [**NZ**, Theorem 2.34, p. 79] and hence is an internal direct product of a subgroup isomorphic to $\mathbb{Z}_p^\bullet$ and a p–group (recall that $\phi(p^{i(p)}) = p^{i(p)-1}(p-1)$). Thus for each integer $i \geq 0$ we have a natural isomorphism

(5) $\qquad (\mathbb{Z}_m^\bullet \times \mathbb{Z}_{n//6t}^\bullet)_{3^i} \cong \prod_p (\mathbb{Z}_p^\bullet)_{3^i} .$

In particular

$$(\mathbb{Z}_m^\bullet \times \mathbb{Z}_{n//6t}^\bullet)_3 \cong \prod_p (\mathbb{Z}_p^\bullet)_3 \cong (\mathbb{Z}_3)^s$$

since $\mathbb{Z}_p^\bullet$ has an element of order 3 if and only if $p \equiv 1 \pmod{3}$ (recall that it is cyclic). Thus $\mathbb{Z}_m^\bullet \times \mathbb{Z}_{n//6t}^\bullet$ has $3^s - 1$ elements of order 3 and hence

$$1 + ((3^s - 1)/2) = (3^s + 1)/2$$

subgroups of order dividing 3 (each subgroup of order 3 contains and is generated by two elements of order 3). Thus if 3 divides n but not t, the set $\mathcal{S}$ has $(3^s + 1)/2$ elements (cf. Remark 7.4). Let us now finish by considering the case that 3 divides t. In this case $\mathbb{Z}_m^\bullet \times \mathbb{Z}_{n//6t}^\bullet \;(\cong \prod_p \mathbb{Z}_{p^{i(p)}}^\bullet)$ has an element of order 3 (namely $(r^{t/3}, 1)$), so some $\mathbb{Z}_p^\bullet$ has an element of order 3. Thus $s > 0$. It suffices to show that if $|\mathcal{S}| \neq 0$, then the condition (4) holds, and that if (4) holds, then $|\mathcal{S}| = 3^{s-1}$. So first suppose $|\mathcal{S}| \neq 0$. Then by Remark 7.4 there exists $a \in \mathbb{Z}_m^\bullet$ with $a^3 = r^{n//3}$. Suppose p is a prime divisor of m with $3_{p-1} \neq 1$ (i.e. $p \equiv 1 \pmod{3}$). We need to show 3_{p-1}, which is the order of the Sylow 3–subgroup of $Z_p^\bullet$, does not equal $|r^{n//3} + p\mathbb{Z}|$. Now this is obvious if $r^{n//3} + p\mathbb{Z}$ is trivial, so suppose otherwise. Then it has order a positive power of 3; since $a^3 = r^{n//3}$, then $3_{p-1} \geq |a + p\mathbb{Z}| = 3|r^{n//3} + p\mathbb{Z}| > |r^{n//3} + p\mathbb{Z}|$, proving (4). Now suppose (4) is valid; we show $|\mathcal{S}| = 3^{s-1}$. Let p be any prime divisor of $mn//6t$. If $3_{p-1} = 1$, then only one element of $\mathbb{Z}_p^\bullet$ has cube equal to $r^{n//3} + p\mathbb{Z} = 1 + p\mathbb{Z}$; if $3_{p-1} > 1$ and p divides $n//6t$, then exactly 3 elements of $\mathbb{Z}_p^\bullet$ have cube equal to 1; and if $3_{p-1} > 1$ and $p \mid m$, then $\mathbb{Z}_p^\bullet$ has exactly three elements b with $b^3 = r^{n//3} + p\mathbb{Z}$ (the hypothesis (4) says $r^{n//3} + p\mathbb{Z}$ does not have maximal 3–power order in the cyclic group $\mathbb{Z}_p^\bullet$). From the isomorphism (5) we now deduce that there are exactly 3^s elements of $\mathbb{Z}_m^\bullet \times \mathbb{Z}_{n//6t}^\bullet$ with cube $(r^{n//3}, 1)$, and hence there are exactly 3^{s-1} cyclic subgroups of $\mathbb{Z}_m^\bullet \times \mathbb{Z}_{n//6t}^\bullet$ such that $\langle (r^{n//3}, 1) \rangle$ is a subgroup of index 3 (each such subgroup is generated by three elements with cube $(r^{n//3}, 1)$). Hence in this case $|\mathcal{S}| = 3^{s-1}$. □

We begin the proof of Theorem 7.3 with a lemma which is the key to the case when 3 does not divide n.

7.6. LEMMA. *Suppose γ and δ are elements of a Frobenius complement of orders p and s, respectively, where p is a prime not dividing s. If $\gamma \delta \gamma^{-1} \in \langle \delta \rangle$, then $\gamma \delta = \delta \gamma$.*

PROOF. There is an integer k with $\gamma \delta \gamma^{-1} = \delta^k$. Since $\delta = \gamma^p \delta \gamma^{-p} = \delta^{k^p}$, then $k^p \equiv 1 \pmod{s}$. Let q be any prime factor of s. Then $\langle \gamma, \delta^{s/q} \rangle$ has order pq and hence is cyclic [**Pa**, Theorem 18.1(ii), p. 194]. Thus $\delta^{s/q} = \gamma \delta^{s/q} \gamma^{-1} = \delta^{ks/q}$. Therefore

$$s/q \equiv ks/q \pmod{s}, \quad \text{so} \quad 1 \equiv k \pmod{q}.$$

Hence $k + s\mathbb{Z}$ is in the kernel of the natural surjective homomorphism $\rho : \mathbb{Z}_s^\bullet \longrightarrow \mathbb{Z}_{s_0}^\bullet$. Therefore the order of $k + s\mathbb{Z}$ in $\mathbb{Z}_s^\bullet$ divides

$$|\ker \rho| = \phi(s)/\phi(s_0),$$

which is a divisor of s (c.f. equation (24) in Chapter 15). Thus $k + s\mathbb{Z}$ has order dividing $(s, p) = 1$. Hence $\gamma \delta = \delta^k \gamma = \delta \gamma$. □

7.7. LEMMA. *Suppose there exists an A_4-complement G with core invariant Δ. Then $n \equiv 2 \pmod{4}$ and G has an r-sequence.*

PROOF. By hypothesis there is a surjective homomorphism $\Upsilon : G \longrightarrow A_4$ whose kernel is the core of G. Then $\Upsilon^{-1}(V_4)$ is a V_4-complement with the same core as G. By Theorem 6.5 it has an r-sequence u, v, x, y. Note vu, v, x, y is also an r-sequence for $\Upsilon^{-1}(V_4)$. Also $n \equiv 2 \pmod{4}$ (apply Theorem 6.5 to $\Upsilon^{-1}(V_4)$). Hence t is odd and 3 does not divide m (Lemma 5.3(A and B1)). Thus $\langle x^{n//3} \rangle$ is a Sylow 3–subgroup of $\Upsilon^{-1}(V_4)$; hence it is a subgroup of index 3 of a (necessarily cyclic) Sylow 3–subgroup $\langle z \rangle$ of G. Since $\langle z^3 \rangle = \langle x^{n//3} \rangle$ we may (and do) assume that z is chosen with $z^3 = x^{n//3}$. $\Upsilon(z)$ has order 3 and hence is a 3-cycle in A_4, and conjugation by $\Upsilon(z)$ must cyclically permute the three nontrivial elements of V_4, namely $\Upsilon(u)$, $\Upsilon(v)$ and $\Upsilon(vu)$. Possibly replacing u by vu we may assume $\Upsilon(zvz^{-1}) = \Upsilon(u)$ and $\Upsilon(zuz^{-1}) = \Upsilon(vu)$. Since $\langle x, y \rangle$ is the kernel of Υ we can therefore write $zvz^{-1} = y^c x^e u$ and $zuz^{-1} = y^d x^f vu$ for some $c, d \in \mathbb{Z}_m$ and $e, f \in \mathbb{Z}_n$. By Lemma 6.8 we can write $zyz^{-1} = y^i$ and $zxz^{-1} = y^j x^g$ where $i \in \mathbb{Z}_m^\bullet$, $j \in \mathbb{Z}_m$, $g \in \mathbb{Z}_n^\bullet$, and $g \equiv 1 \pmod{t}$.

We next prove that u and v commute with x and y, so the invariant and reduced invariant of the r-sequence u, v, x, y of the V_4-complement $\Upsilon^{-1}(V_4)$ are trivial. We can write $uyu^{-1} = y^a$ and $vyv^{-1} = y^b$ for $a, b \in \mathbb{Z}_m^\bullet$. Conjugating the first equation by z yields the equation

$$y^{ai} = y^d x^f v u y^i u^{-1} v^{-1} x^{-f} y^{-d} = y^{iabr^f}.$$

Hence $br^f = 1$. Now b has order dividing 2 and r^f odd order, so $b = 1$ and $r^f = 1$, so $f \equiv 0 \pmod{t}$. Similarly, conjugating the second equation by z shows that $b = ar^e$, so $r^e = 1$ (whence $e \equiv 0 \pmod{t}$) and $a = 1$. Now write $uxu^{-1} = x^h$ and $vxv^{-1} = x^k$ where $h, k \in \mathbb{Z}_n^\bullet$ and $h \equiv k \equiv 1 \pmod{t}$ (Lemma 6.8). Conjugating the equation $ux = x^h u$ by z yields that

$$y^{d+j}x^{f+ghk}vu = y^d x^f vu y^j x^g = (y^j x^g)^h y^d x^f vu = y^{j+dr}x^{gh+f}vu$$

(using Lemma 6.7 and the facts that $r^f = 1$ and $g \equiv h \equiv 1 \pmod{t}$). Hence $k = 1$ and $d = dr$. But then $d = 0$ since $(r-1, m) = 1$. Similarly conjugating $vx = x^k v$ by z shows that $c = cr$ and $gk = gh$ (using the fact that $k \equiv 1 \pmod{t}$ and $r^e = 1$), so $c = 0$ and $h = k = 1$. This completes the argument that u and v commute with x and y.

Let us summarize our partially simplified relations between u, v, x, y, z: u, v, x, y is an r-sequence for $\Upsilon^{-1}(V_4)$ with trivial invariant, $z^3 = x^{n//3}$, $zuz^{-1} = x^f vu$, $zvz^{-1} = x^e u$, $zxz^{-1} = y^j x^g$, and $zyz^{-1} = y^i$. We next argue that u, v, x, y, z could have been chosen so that $j = 0$. Note that $y^{i^3} = z^3 y z^{-3} = x^{n//3} y x^{-n//3} = y^{r^{n//3}}$, so $i^3 = r^{n//3}$. Let $I = 1 + i + i^2 + \cdots + i^{3(3_n)-1}$, so

$$(i-1)I = i^{3(3_n)} - 1 = \left(r^{n//3}\right)^{(3_n)} - 1 = r^n - 1 = 0.$$

By Lemma 6.8 (applied with $s = 3$ to conjugation by z on $\langle x, y \rangle$) we have that $g^3 = 1$ and $(1 + i + i^2)j \equiv 0 \pmod{m}$, and (applied with $s = 3(3_n)$, so that $z^s = \left(x^{n//3}\right)^{3_n} = 1$ trivially commutes with x and y) we have that m is the product of the relatively prime factors $(m, i-1)$ and (m, I). Since $1 + i + i^2$ divides I, then $(m, 1 + i + i^2)$ divides (m, I) and hence $j \equiv 0 \pmod{m/(m, I)}$, i.e., $j \equiv 0 \pmod{(m, i-1)}$. Now $(i-1)/(m, i-1)$ is relatively prime to $m/(m, i-1) = (m, I)$, so

there exists $i^* \in \mathbb{Z}$ with
$$i^*(i-1)/(m, i-1) \equiv -j/(m, i-1) \pmod{(m, I)}$$
and hence
$$i^*(i-1) \equiv -j \pmod{m}.$$
Let $\check{x} = y^{i^*}x$. Routine computations show $u, v, \check{x}, y$ is an r-sequence for $\Upsilon^{-1}(V_4)$ with trivial invariant. (E.g. use Lemma 6.7 to verify that $\check{x}^{n//2} = x^{n//2} = v^2$.) Further,
$$\check{x}^{n//3} = (y^{i^*}x)^{n//3} = y^{i^*(1+r+r^2+\cdots+r^{(n//3)-1})}x^{n//3} = x^{n//3} = z^3$$
because $(r-1, m) = 1$ and
$$(r-1)i^*(1+r+\cdots+r^{(n//3)-1}) \equiv (r^{n//3}-1)i^* \equiv (i^3-1)i^*$$
$$\equiv i^*(i-1)(1+i+i^2) \equiv -j(1+i+i^2) \equiv 0 \pmod{m}.$$

Next
$$z\check{x}z^{-1} = zy^{i^*}xz^{-1} = y^{ii^*}y^jx^g = y^{i^*}x^g = \check{x}^g$$
because
$$(ii^* + j) - i^* = i^*(i-1) + j = -j + j \equiv 0 \pmod{m}.$$

Further since $e \equiv f \equiv 0 \pmod{t}$ as shown earlier, then $zuz^{-1} = \check{x}^f vu$ and $zvz^{-1} = \check{x}^e u$ by Lemma 6.7. Thus we may without loss of generality assume $j = 0$ (replace x by $\check{x}$); none of the relations reviewed at the beginning of this paragraph are changed.

Conjugating both sides of the equation $u^2 = x^{n/2}$ by z we obtain
$$x^{n/2} = x^f vu x^f vu = x^{2f}(vu)^2 = x^{2f}x^{n/2}$$
so $2f = 0$. Similarly since $v^2 = x^{n//2}$, then
$$x^{g(n//2)} = x^e u x^e u = x^{2e}x^{n/2},$$
so $g(n//2) \equiv 2e + (n/2) \pmod{n}$. Since $n \equiv 2 \pmod{4}$, therefore $n//2 = n/2$, and so $2e = 0$. Hence $e \equiv f \equiv 0 \pmod{n/2}$, so $x^e, x^f \in \langle u^2 \rangle$. Let $\check{v} = x^f v$ and $\check{u} = x^{e+f}u$. Then $\check{v}^2 = x^{n//2}$; $\check{v}$ and $\check{u}$ commute with x and y;
$$z\check{u}z^{-1} = x^{e+f}x^f vu = \check{v}\check{u}; \quad \text{and} \quad z\check{v}z^{-1} = x^f x^e u = \check{u}.$$
Then $\check{u}, \check{v}, x, y, z$ is an r-sequence for G. $\square$

7.8. LEMMA. *Suppose an A_4-complement with core invariant Δ has an r-sequence with invariant (a, g). Then $g^3 = 1$, $a^3 = r^{n//3}$, $g \equiv 1 \pmod{(6t)_n}$, and the reduced invariant $\langle (a, g + n//6t\mathbb{Z}) \rangle$ is in $\mathcal{S}$.*

PROOF. Let u, v, x, y, z be an r-sequence with invariant (a, g). Since
$$y^{a^3} = z^3 y z^{-3} = x^{n//3} y x^{-n//3} = y^{r^{n//3}}$$
then $a^3 = r^{n//3}$. Lemma 6.8 (applied to conjugation by z, with $s = 3$) implies $g^3 = 1$ and $g \equiv 1 \pmod{t}$. Since z commutes with $z^3 = x^{n//3}$, then $x^{n//3} = zx^{n//3}z^{-1} = x^{g(n//3)}$, so $g \equiv 1 \pmod{3_n}$ (recall that $n = (n//3)(3_n)$). Since $n \equiv 2 \pmod{4}$ (by the previous lemma) and $g^3 = 1$, then $g \equiv 1 \pmod{2_n}$. By Lemma 6.9 (applied

with $s = 3$ and $k = t//3$), $g \equiv 1 \pmod{(t//3)_n}$. Thus $g \equiv 1 \pmod{(6t)_n}$ (every prime dividing $6t$ divides 2 or 3 or $t//3$). Now $\langle r^{n//3} \rangle$ is the Sylow 3-subgroup of T and $\langle (a, g + n//6t\mathbb{Z})^3 \rangle = \langle (r^{n//3}, 1) \rangle$, so $T^* \times \langle 1 \rangle$ has index 1 or 3 in the reduced invariant. If $3 | t$, then the index is $3 = (3, n) = (3, t)$. If $3 | n$ but $3 \nmid t$, then the index is indeed either $1 = (3, t)$ or $3 = (3, n)$. Finally suppose 3 does not divide n. Then $z^3 = x^{n//3} = x^n = 1$, so $|z| = 3$ is not a factor of $|x|$ or $|y|$. Hence by Lemma 7.6, z commutes with both x and y, so the invariant (a, g) is trivial. Thus the reduced invariant has index $1 = (3, t) = (3, n)$ over the (trivial) Sylow 3-subgroup of T. Therefore in all cases the reduced invariant is in $\mathcal{S}$. □

7.9. LEMMA. *Suppose u, v, x, y, z and u_1, v_1, x_1, y_1, z_1 are r-sequences with the same invariant (a, g) for A_4-complements G and G_1 with core invariant Δ, respectively. Then there is an isomorphism from G to G_1 taking u, v, x, y, and z to u_1, v_1, x_1, y_1, and z_1, respectively.*

PROOF. The lemma follows from the fact that Δ and (a, g) completely determine a presentation of G by generators and relations with u, v, x, y, z as the generators. □

We now introduce notations that will be used in Chapters 9 and 10 as well as in the next theorem.

7.10. NOTATION. Suppose $n \equiv 2 \pmod 4$. Let $\sigma \in \text{Aut}\mathbb{Z}[\zeta_{mn/2t}]$ fix $\zeta_{n/2t}$ and map ζ_m to ζ_m^r. Then let $\mathcal{A}_0 = (\mathbb{Z}[\zeta_{mn/2t}], \sigma, t, \zeta_{n/2t})$ and let $J = \langle \zeta_m, \hat{\sigma} \rangle$ (a subgroup of $\mathcal{A}_0^\bullet$). If $n \neq 2$ then $(m, n/2, T)$ is a proper Frobenius triple (Lemma 5.3B(3)) and J and $\mathcal{A}_0$ are the 1-complement and truncated group ring constructed as in Theorem 5.2C with respect to this triple. If $n = 2$, then $J = \langle 1 \rangle$ and we may identify $\mathcal{A}_0$ with $\mathbb{Z} \cong \mathbb{Z}\langle J \rangle$. (Note that since $2_{mn} = 2$, then $\mathbb{Z}[\zeta_{mn/2t}] = \mathbb{Z}[\zeta_{mn/t}]$, so the σ above is the same as the σ of Theorem 5.2C, but the $\hat{\sigma}$ is not!). We let x_0, y_0 be an r-sequence for J if $n \neq 2$ and set $x_0 = y_0 = 1$ if $n = 2$.

The next theorem uses the above notation and the group H_{24} of Example 3.1. Together with Lemmas 7.7 and 7.8 it completes the proof of Theorem 7.3 in the case that n is not divisible by 3.

7.11. THEOREM. *Suppose that $n \equiv 2 \pmod 4$ and that 3 does not divide n. Then $G := H_{24} \times J$ is the unique (up to isomorphism) A_4-complement with core invariant Δ, and the natural map $G \longrightarrow \mathbb{Z}[H_{24}] \otimes \mathcal{A}_0$ induces an isomorphism $\mathbb{Z}\langle G \rangle \longrightarrow \mathbb{Z}[H_{24}] \otimes \mathcal{A}_0$. $\mathbb{Z}\langle G \rangle$ is free of rank $4\phi(mn)$ as a $\mathbb{Z}$-module. $\mathbb{Q}\langle G \rangle$ is simple with degree $2t$ and with dimension $4\phi(mn)$ over $\mathbb{Q}$; its center is naturally isomorphic to the center of $\mathbb{Q}\mathcal{A}_0$ (namely, $\mathbb{Q}[\zeta_{mn/t}]^\sigma$). Finally, $(\mathbf{i}, 1)$, $(\mathbf{k}, 1)$, $(-1, \hat{\sigma})$, $(1, \zeta_m)$, $(\boldsymbol{\alpha}^2, 1)$ is an r-sequence for G.*

PROOF. By Theorem 5.2 (and Notation 7.10) the inclusion $J \longrightarrow \mathcal{A}_0$ induces an isomorphism $\mathbb{Z}\langle J \rangle \longrightarrow \mathcal{A}_0$. By Theorem 3.2 the inclusion $H_{24} \longrightarrow \mathbb{Z}[H_{24}]$ induces an isomorphism $\mathbb{Z}\langle H_{24} \rangle \longrightarrow \mathbb{Z}[H_{24}]$. Now H_{24} and J have relatively prime orders by hypothesis (and Lemma 5.3(B1)), so by the Direct Product Lemma 2.9 G is a Frobenius complement and the natural map $G \longrightarrow \mathbb{Z}[H_{24}] \otimes \mathcal{A}_0$ induces an isomorphism $\mathbb{Z}\langle G \rangle \longrightarrow \mathbb{Z}[H_{24}] \otimes \mathcal{A}_0$. The Hurwitz ring $\mathbb{Z}[H_{24}] = \mathbb{Z}[\mathbf{i}, \mathbf{j}, \boldsymbol{\alpha}]$ is a free $\mathbb{Z}$-module by Lemma 3.5. Its rank must equal the rational dimension of the ring $\mathbb{Q} \otimes \mathbb{Z}[H_{24}] \cong \mathbb{Q}[\mathbf{i}, \mathbf{j}]$, which is a central simple quaternion algebra over $\mathbb{Q}$ of rational dimension 4. By Theorem 5.2 $\mathcal{A}_0$ is free of rank $\phi(mn)$ as a $\mathbb{Z}$-module, and $\mathbb{Q}\mathcal{A}_0$

is simple with dimension $\phi(mn)$ over $\mathbb{Q}$ and with center $\mathbb{Q}[\zeta_{mn/t}]^\sigma$. Thus $\mathbb{Z}\langle G \rangle$ is free of rank $4\phi(mn)$ as a $\mathbb{Z}$-module, $\mathbb{Q}\langle G \rangle$ is simple with dimension $4\phi(mn)$ over $\mathbb{Q}$, and the center of $\mathbb{Q}\langle G \rangle$ is naturally isomorphic to the center of $\mathbb{Q}\mathcal{A}_0$ [**P**, Lemma b(ii), p. 225 and Lemma c(ii), p. 226]. The degree of $\mathbb{Q}\langle G \rangle$ is the product of the degrees of $\mathbb{Q}[\mathbf{i}, \mathbf{j}]$ and $\mathbb{Q}\mathcal{A}_0$, namely, $2t$. Now let $x = (-1, \widehat{\sigma})$, $y = (1, \zeta_m)$, $u = (\mathbf{i}, 1)$, $v = (\mathbf{k}, 1)$ and $z = (\boldsymbol{\alpha}^2, 1)$. Routine computations show that $G = \langle u, v, x, y, z \rangle$; that $\langle x, y \rangle$ is a normal subgroup of G with $G/\langle x, y \rangle \cong H_{24}/\langle -1 \rangle \cong A_4$; and that $\langle x, y \rangle$ is a 1-complement with invariant (m, n, T). Thus G is an A_4-complement with core invariant Δ. Using Lemma 3.4 one checks that the sequence u, v, x, y, z is an r-sequence for G. Now suppose H is any A_4-complement with core invariant Δ. By Lemmas 7.7 and 7.8 H has an r-sequence, say with invariant (a, g); $g \equiv 1$ (mod $(6t)_n$); and the reduced invariant $\langle (a, g + n//6t\mathbb{Z}) \rangle$ is in $\mathcal{S}$. But the only element of $\mathcal{S}$ is the trivial group. Thus (a, g) is trivial (recall that $n = (6t)_n(n//6t)$). Therefore every A_4-complement with core invariant Δ has trivial invariant and hence, by Lemma 7.9, they are all isomorphic. □

7.12. REMARK AND NOTATION. Suppose $n \equiv 2 \pmod{4}$. Let $H = \langle u, v, x, y \rangle$ be the V_4-complement with core invariant Δ and trivial reduced invariant constructed in Theorem 6.14 as a subgroup of the algebra $\mathcal{B}$ (and denoted in the statement of Theorem 6.14 by G). In the next theorem we will use H and $\mathcal{B}$ to construct A_4-complements with core invariant Δ and with reduced invariant any given element in $\mathcal{S}$ in the case that 3 divides n.

We remark on some alternative approaches to A_4-complements. First, it is not hard to show that $H \cong D_8 \times J$ and hence that $\mathcal{B} \cong \mathbb{Z}[\mathbf{i}, \mathbf{j}] \otimes \mathcal{A}_0$ where J and $\mathcal{A}_0$ are as in Notation 7.10. Thus in the next theorem H and $\mathcal{B}$ could have been replaced by $D_8 \times J$ and $\mathbb{Z}[\mathbf{i}, \mathbf{j}] \otimes \mathcal{A}_0$, respectively. Next suppose that 3 does not divide n. Let $A = -(1 + u + v + uv)/2 \in \mathbb{Q}\mathcal{B}$. Then the subgroup $\langle x, y, u, v, A \rangle$ of $\mathbb{Q}\mathcal{B}$ is an A_4-complement with core invariant Δ. This observation gives an another approach to the material of the previous theorem.

7.13. THEOREM. *Suppose that $n \equiv 2 \pmod{4}$, 3 divides n, and $S \in \mathcal{S}$.*

(A) There exists $a \in \mathbb{Z}_m^\bullet$ and $g \in \mathbb{Z}_n^\bullet$ with $S = \langle (a, g + n//6t\mathbb{Z}) \rangle$, $g \equiv 1$ (mod $(6t)_n$), and $a^3 = r^{n//3}$.

(B) There is an automorphism τ of $\mathcal{B}$ which restricts on H to a group automorphism fixing $x^{n//3}$ and mapping u, v, x, and y to vu, u, x^g, and y^a, respectively. Moreover, $\tau^3(w) = x^{n//3} w x^{-n//3}$ for all $w \in H$. Let $\mathcal{C} = (\mathcal{B}, \tau, 3, x^{n//3})$, $z = \widehat{\tau}$, and $G = \langle u, v, x, y, z \rangle$. Then G is an A_4-complement with core invariant Δ and reduced invariant S; and u, v, x, y, z is an r-sequence for G with invariant (a, g).

(C) The inclusion $G \longrightarrow \mathcal{C}$ induces an isomorphism $\theta : \mathbb{Z}\langle G \rangle \longrightarrow \mathcal{C}$. $\mathbb{Z}\langle G \rangle$ is free of rank $12\phi(mn)$ as a $\mathbb{Z}$-module. θ extends to an isomorphism $\mathbb{Q}\langle G \rangle \longrightarrow \mathbb{Q}\mathcal{C}$. $\mathbb{Q}\mathcal{C}$ is a simple algebra with dimension $12\phi(mn)$ over $\mathbb{Q}$, with degree $2t(|S|, 3)$, and with center K^τ if $|S| \neq 1$ and $K[z(1 - u - v - vu)]$ if $|S| = 1$, where K denotes the center of $\mathbb{Q}\mathcal{B}$.

The group H and ring $\mathcal{B}$ of the above theorem are those of the preceding remark. The assertions about τ in (B) guarantee that $\mathcal{C}$ is properly defined (cf. Notation 2.6B). By Theorem 6.15 the center K of $\mathbb{Q}\mathcal{B}$ is the field $\mathbb{Q}[\zeta_{mn/t}]^\sigma$ of Theorem 5.2; if $|S| = 1$ we will see that $-z(1 - u - v - vu)/2$ is a root of unity of order $3(3_n)$ and that it commutes with v, so that in this case it generates cubic extensions both of L and of $L[v]$.

PROOF. $S = \langle (a, g_0) \rangle$ for some $a \in \mathbb{Z}_m^\bullet$ and $g_0 \in \mathbb{Z}_{n//6t}^\bullet$ where $a^3 = r^{n//3}$ and $g_0^3 = 1$ (Remark 7.4; note that $r^{n//3} = 1$ if 3 does not divide t). By the Chinese Remainder Theorem there exists $g \in \mathbb{Z}_n^\bullet$ with $g \equiv g_0 \pmod{n//6t}$ and $g \equiv 1 \pmod{(6t)_n}$, so that $g^3 \equiv 1 \pmod{n}$ and $g \equiv 1 \pmod{t}$.

Since u and v commute with x and y, then vu and u commute with x^g and y^a. Since $g \equiv 1 \pmod{t}$, $x^g y^a x^{-g} = (y^a)^r$. Also $|x^g| = n$ and $|y^a| = m$ and $\langle x^g, y^a \rangle$ is the core of H. Finally, $\langle vu, u \rangle$ is a Sylow 2–subgroup of H, and $(vu)^2 = (x^g)^{n//2}$ since $n \equiv 2 \pmod{4}$. Hence vu, u, x^g, y^a is an r–sequence for H with the same trivial invariant as u, v, x, y. Thus there is an automorphism τ_0 of H taking u, v, x and y to vu, u, x^g and y^a, respectively (argue as in the proof of Theorem 7.9). Hence there is an automorphism τ_1 of $\mathbb{Z}\langle H \rangle$ taking $\bar{u}, \bar{v}, \bar{x}, \bar{y}$ to $\overline{vu}, \bar{u}, \bar{x}^g, \bar{y}^a$ and therefore (since the inclusion $H \longrightarrow \mathcal{B}$ induces the isomorphism $\mathbb{Z}\langle H \rangle \longrightarrow \mathcal{B}$) an automorphism τ of $\mathcal{B}$ restricting to τ_0 on H. Since $a^3 = r^{n//3}$ and $g^3 \equiv 1$, then $\tau^3(x) = x^{g^3} = x = x^{n//3} x x^{-n//3}$ and $\tau^3(y) = y^{a^3} = y^{r^{n//3}} = x^{n//3} y x^{-n//3}$; the analogous formulas with y replaced by u or v hold since u and v commute with x. Thus $\tau^3(w) = x^{n//3} w x^{-n//3}$ for all $w \in \mathcal{B}$. Finally, $\tau(x^{n//3}) = x^{g(n//3)} = x^{n//3}$ since $g \equiv 1 \pmod{(6t)_n}$ (so $g \equiv 1 \pmod{3_n}$ and $(n//3)g \equiv n//3 \pmod{n}$). This completes the proof of the existence of τ. Since $\mathcal{B}$ is a free $\mathbb{Z}$–module of rank $4\phi(mn)$, then $\mathcal{C}$ is by construction a free $\mathbb{Z}$–module of rank $12\phi(mn)$. The Cyclic Extension Lemma 2.8 says that G is a Frobenius group and that the inclusion $G \longrightarrow \mathcal{C}$ induces an isomorphism $Z\langle G \rangle \longrightarrow \mathcal{C}$. (We use here the fact that the inclusion $H \longrightarrow \mathcal{B}$ induces an isomorphism $\mathbb{Z}\langle H \rangle \longrightarrow \mathcal{B}$, cf. Theorem 6.14.) G is easily checked to be an A_4–complement with core $\langle x, y \rangle$ and hence with core invariant Δ. By construction u, v, x, y, z is an r–sequence for G with invariant (a, g) and reduced invariant S. Also $\mathbb{Q}\langle G \rangle \cong \mathbb{Q} \otimes \mathcal{C}$ has dimension $12\phi(mn)$ as a $\mathbb{Q}$–algebra. We can identify $\mathbb{Q}\mathcal{C} = \mathbb{Q} \otimes \mathcal{C}$ with $(\mathbb{Q}\mathcal{B}, \tau, 3, x^{n//3})$ (cf. Remark 2.6C). It remains to argue that $\mathbb{Q}\mathcal{C}$ is simple and to compute its center and degree.

First suppose that S is nontrivial. If τ were trivial on $\langle x^t, y \rangle$, then $y = \tau(y) = y^a$ so $a = 1$, and $x^t = \tau(x^t) = x^{tg}$ so $g \equiv 1 \pmod{n/t}$. Thus $g = 1$ since by construction $g \equiv 1 \pmod{t_n}$. This contradicts the hypothesis that S is nontrivial. Hence τ is nontrivial on the cyclic group $\langle x^t, y \rangle$ and hence on the cyclotomic extension $L = \mathbb{Q}[x^t, y] = \mathbb{Q}[\zeta_{nm/t}]$. Both σ and τ induce automorphisms on L (and hence on all of the subfields of L). On L we have $\tau^3 = \sigma^{n//3}$ and $\tau\sigma = \sigma\tau$, so $|\langle \sigma|L, \tau|L \rangle| \le 3t$. Suppose $\tau|L = (\sigma|L)^i$ for some integer i. Then since $\sigma(x^t) = x^t$, we have $x^{gt} = \tau(x^t) = x^t$, so $g \equiv 1 \pmod{n/t}$, whence $g \equiv 1 \pmod{n//6t}$. Thus $a \ne 1$ (by hypothesis S is nontrivial). Also $y^{r^i} = \sigma^i(y) = \tau(y) = y^a$, so $a = r^i$. Thus $|a|$ divides t. Since $a^3 = r^{n//3}$, the order of a is a power of 3. Hence 3 divides t. Since $r^{3i} = a^3 = r^{n//3}$, then $3i \equiv n//3 \pmod{t}$. But then $0 \equiv n//3 \pmod{3}$, a contradiction. Hence $\tau|L \notin \langle \sigma|L \rangle$. Thus $|\langle \sigma|L, \tau|L \rangle| = 3t$. Setting $K = L^\sigma$ we therefore have

$$\begin{aligned} [K : K^\tau] &= [L : K^\tau][L : K]^{-1} = [L : L^{\langle \sigma, \tau \rangle}] t^{-1} \\ &= |\langle \sigma|L, \tau|L \rangle| t^{-1} = 3. \end{aligned}$$

Thus by the last part of Theorem 2.5 $\mathbb{Q}\mathcal{C}$ is simple with center K^τ. Hence the degree of $\mathbb{Q}\mathcal{C}$ is triple the degree of $\mathbb{Q}\mathcal{B}$, namely $6t = 2t(|S|, 3)$ (cf. Theorem 6.15).

Next consider the case that S is trivial. Then z commutes with x and y. Now x and y also commute with u and v, and hence with $A := (-1 + u + v + vu)/2$ (an element of $\mathcal{QC}$). Also A and z commute since conjugation by z just permutes the summands of A. Hence $|Az| = |z|$ (the order of z is a nontrivial power of 3 since $z^3 = x^{n//3}$ by construction, and $|A| = 3$ by Lemma 3.4E). (In the application of Lemma 3.4E we are identifying u, v, and A with **i**, **k**, and $\boldsymbol{\alpha}$. Formally we have a homomorphism $\langle \mathbf{i}, \mathbf{k} \rangle \longrightarrow G$ taking **i** and **k** to u and v, and this map induces a homomorphism $\rho : \mathbb{Z}[\mathbf{i}, \mathbf{k}] \longrightarrow \mathcal{C}$ taking $2\boldsymbol{\alpha}$ to $2A$; ρ can be used to push the formulas of Lemma 3.4 into $\mathcal{C}$.) Further, Az commutes not only with x and y but also with u and v, since conjugation by A and by z give inverse automorphisms of $\langle u, v \rangle$ (Lemma 3.4A). Hence Az is in the center of $\mathcal{QC}$. Thus the center of $\mathcal{QC}$ contains $K[Az]$. Since $a = 1$, then $r^{n//3} = 1$, so t divides $n//3$. Hence $x^{n//3} \in K$. Now Az is a root of the polynomial $X^3 - x^{n//3} \in K[X]$. The roots of this polynomial are roots of unity of order $3(3_n)$ and hence do not lie in K (recall $K \subset \mathbb{Q}[\zeta_{mn/t}]$). Thus $K[Az]$ is a cubic field extension of K. Hence $\mathcal{QC} \cong \mathcal{QB} \otimes_K K[Az]$ [**P**, Proposition a, p. 225]. Thus $\mathcal{QC}$ is simple with center $K[Az]$ [**P**, Proposition b(ii), p. 226]. The degree of $\mathcal{QC}$ is $2t = 2t(|S|, 3)$ because $\mathcal{QB}$ has degree $2t$ (Theorem 6.15), $[\mathcal{QC} : \mathcal{QB}] = 3$, and $[K[Az] : K] = 3$. $\square$

It remains to solve the isomorphism problem when 3 divides n, i.e., to show that in this case every A_4–complement with core invariant Δ has a unique reduced invariant and that the reduced invariant determines the A_4–complement up to isomorphism.

7.14. LEMMA. *An A_4–complement G with core invariant Δ has only one reduced invariant.*

PROOF. By Theorem 7.11 we may suppose 3 divides n. Let u, v, x, y, z and u_1, v_1, x_1, y_1, z_1 be r–sequences for G with invariants (a, g) and (a_1, g_1), respectively. By Lemma 7.8
$$a^3 = a_1^3 = r^{n//3}, \quad g^3 = g_1^3 = 1, \quad g \equiv g_1 \pmod{(6t)_n}.$$
We can write $y_1 = y^b$ and $x_1 = y^c x^h$ for some $b \in \mathbb{Z}_m^\bullet$, $c \in \mathbb{Z}_m$, and $h \in \mathbb{Z}_n^\bullet$. As usual $h \equiv 1 \pmod{t}$ since
$$y^{br^h} = x_1 y_1 x_1^{-1} = y_1^r = y^{br}.$$
We can also write $z_1 = y^d x^f v^i u^j z^s$ where $s = 1$ or 2, $d \in \mathbb{Z}_m$, $f \in \mathbb{Z}_n$, $i, j \in \mathbb{Z}$. Then
$$y^{a_1 b} = y_1^{a_1} = z_1 y_1 z_1^{-1} = y^{ba^s r^f},$$
so $a_1 = a^s r^f$ and hence $r^{n//3} = r^{s(n//3)+3f}$. Therefore
$$(6) \qquad (s-1)(n//3) + 3f \equiv 0 \pmod{t}.$$
Similarly modulo the normal subgroup $\langle y \rangle$ we have
$$x^{g_1 h} \equiv x_1^{g_1} \equiv z_1 x_1 z_1^{-1} \equiv x^{h g^s}$$
and so $g_1 = g^s$. If 3 does not divide t, then the congruence (6) implies that $3f \equiv 0 \pmod{t}$, and so $f \equiv 0 \pmod{t}$. Hence $a_1 = a^s$. Thus
$$(a, g + n//6t\mathbb{Z})^s = (a_1, g_1 + n//6t\mathbb{Z}),$$
so the reduced invariants of the two r–sequences are the same. Now suppose 3 divides t. Then the congruence (6) says $(s-1)(n//3) \equiv 0 \pmod 3$, so $s = 1$. Thus $g = g_1$ and $f \equiv 0 \pmod{t/3}$. It follows that r^f has order dividing 3 and hence lies in

the unique subgroup of $\langle r \rangle = T$ of order 3, namely $\langle r^{t/3} \rangle$. Thus $(a, g + n/\!/6t\mathbb{Z})$ and $(a_1, g_1 + n/\!/6t\mathbb{Z})$ lie in the same coset of $(\mathbb{Z}_m^\bullet \times \mathbb{Z}_{n/\!/(6t)}^\bullet)/\langle (r^{t/3}, 1) \rangle$ and so generate the same subgroup of $\mathbb{Z}_m^\bullet \times \mathbb{Z}_{n/\!/(6t)}^\bullet$ (cf. Remark 7.4). □

The next lemma will complete the proof of Theorem 7.2.

7.15. LEMMA. *Suppose G and G_1 are A_4-complements with core invariant Δ and with the same reduced invariant. Then $G \cong G_1$.*

PROOF. We may suppose 3 divides n (Theorem 7.11). By hypothesis there are r-sequences u, v, x, y, z and u_1, v_1, x_1, y_1, z_1 for G and G_1 with invariants (a, g) and (a_1, g_1), respectively, such that

$$S := \langle (a, g + n/\!/6t\mathbb{Z}) \rangle = \langle (a_1, g_1 + n/\!/6t\mathbb{Z}) \rangle.$$

Our strategy is to construct a new r-sequence for G with invariant (a_1, g_1) and then to invoke Lemma 7.9 to conclude that $G \cong G_1$.

First suppose 3 does not divide t, so that the reduced invariant S is a subgroup of $\mathbb{Z}_m^\bullet \times \mathbb{Z}_{n/\!/6t}^\bullet$ of order dividing 3 (Remark 7.4). Then for s equal to either 1 or 2,

$$(a, g + n/\!/6t\mathbb{Z})^s = (a_1, g_1 + n/\!/6t\mathbb{Z}).$$

Hence $a^s = a_1$ and $g^s \equiv g_1$ (apply Lemma 7.8 which says $g^s \equiv g_1 \pmod{(6t)_n}$). If $s = 1$ there is nothing to prove, so suppose $s = 2$. Consider the sequence v^3, u^3, x^i, y, z^2 where $i \in \mathbb{Z}$ is chosen so that $i \equiv 2 \pmod{3_n}$ and $i \equiv 1 \pmod{n/\!/3}$. Since 3 does not divide t, then $i \equiv 1 \pmod{t}$, so $x^i y x^{-i} = y^r$. Also $|x^i| = n$ since $(i, n) = 1$. Thus v^3, u^3, x^i, y is an r-sequence for the V_4-complement $\langle u, v, x, y \rangle$ with trivial invariant (cf. Definition 6.4). Further, $(z^2)^3 = x^{2(n/\!/3)} = (x^i)^{n/\!/3}$ since $i \equiv 2 \pmod{3_n}$. Also $z^2 y z^{-2} = y^{a^2} = y^{a_1}$ and $z^2 x^i z^{-2} = x^{ig^2} = (x^i)^{g_1}$. One easily computes that $z^2 u^3 z^{-2} = v^3$ and $z^2 v^3 z^{-2} = (vu)^3 = u^3 v^3$. It follows that v^3, u^3, x^i, y, z^2 is an r-sequence for G with invariant (a_1, g_1), so $G \cong G_1$.

Now suppose 3 divides t. In this case the construction of a suitable r-sequence for G is a bit more subtle. Pick an integer s' with $s' \equiv 1 \pmod{3_t}$ and $s' \equiv 0 \pmod{n/\!/3}$, and set $s = s't/3$. Then $s \equiv t/3 \pmod{t}$ and $s \equiv 0 \pmod{(n/\!/3)(3_t/3)}$. Since 3 divides t and $g \equiv 1 \pmod{t}$, then $1 + g + g^2 \equiv 0 \pmod{3}$. Hence there exists $\gamma \in \mathbb{Z}_n$ with $3\gamma = 1 + g + g^2$. Since $(n/\!/3, 3_n) = 1$, there exists $\mu \in \mathbb{Z}$ with $\mu(n/\!/3) \equiv 1 \pmod{3_n}$. By the Chinese Remainder Theorem we can pick $h \in \mathbb{Z}_n$ with $h \equiv 0 \pmod{n/\!/6t}$, $h \equiv 0 \pmod{2_n}$, and $h \equiv \mu s' \gamma \pmod{3_n}$. Now set $\tau = 1 + th \in \mathbb{Z}_n$, so $\tau \equiv 1 \pmod{t}$ and $(\tau, n) = 1$. By the choices of γ, μ, and h (respectively) we have that modulo 3_n

$$s'(t/3)(1 + g + g^2) \equiv s'(t/3)(3\gamma) \equiv t\mu s'\gamma(n/\!/3) \equiv th(n/\!/3)$$

and hence $s(1 + g + g^2) \equiv (\tau - 1)(n/\!/3) \pmod{3_n}$. Since $s \equiv 0 \pmod{n/\!/3}$ we conclude that

(7) $$(\tau - 1)(n/\!/3) \equiv s(1 + g + g^2) \pmod{n}.$$

Now consider the sequence u, v, x^τ, y, $x^s z$. We show this is an r-sequence for G. Clearly u, v, x^τ, y is an r-sequence for $\langle u, v, x, y \rangle$ with trivial invariant (one uses here only the facts that $\tau \equiv 1 \pmod{t}$ and $(\tau, n) = 1$). Next $(x^s z)^3 = x^{s(1+g+g^2)} x^{n/\!/3} = (x^\tau)^{n/\!/3}$ (by the congruence (7)), $(x^s z) u (x^s z)^{-1} = vu$, and $(x^sz)v(x^sz)^{-1} = u$ (since u and v commute with x). Finally, $(x^sz)y(x^sz)^{-1} = y^{ar^s} = y^{ar^{t/3}}$ and $(x^sz)x^\tau(x^sz)^{-1} = (x^\tau)^g$. This shows u, v, x^τ, y, x^sz is an r–sequence for G with invariant $(ar^{t/3}, g)$. By Remark 7.4, $a_1 = a(r^{t/3})^i$ for $i = 0$, 1 or 2 and $g = g_1$. If $i = 0$ then $a_1 = a$, so G and G_1 have r-sequences with the same invariant; if $i = 1$ then G and G_1 still have r-sequences with the same invariant (a_1, g_1): the sequence u, v, x^τ, y, x^sz works for G. If $i = 2$ then the above argument shows u, v, x^{τ^2}, y, $x^{s(\tau+1)}z$ is an r-sequence for G with invariant (a_1, g_1) (repeat the construction of the r-sequence u, v, x^τ, y, x^sz from the r-sequence u, v, x, y, z but now starting with the r-sequence u, v, x^τ, y, x^sz) . Hence for any value of i, $G \cong G_1$. $\square$

CHAPTER 8

Frobenius complements with core index 24

The next definition focuses on a type of presentation by generators and relations of S_4–complements. We continue of course to use the notation of Notation 6.2.

8.1. DEFINITION. Suppose G is an S_4–complement with core invariant Δ. Suppose u, v, w, x, y, z is a sequence of elements of G and (a,g) is an element of $\mathbb{Z}_m^\bullet \times \mathbb{Z}_n^\bullet$ such that $H := \langle u, v, x, y, z \rangle$ is an A_4–complement with the same core as G; that u, v, x, y, z is an r–sequence for H with trivial invariant; and that $w^2 = u$, $wxw^{-1} = x^g$, $wyw^{-1} = y^a$, and $wzw^{-1} = x^{-n//3}vz^2$. We then call u, v, w, x, y, z an r–sequence for G with invariant (a,g) and reduced invariant $\langle(a, g + n//6t\mathbb{Z})\rangle$.

One assertion of the next theorem is that an S_4–complement has an r–sequence and all such have the same reduced invariant.

8.2. THEOREM. *There exists an S_4–complement with core invariant Δ if and only if 3 does not divide t and $n \equiv 2 \pmod 4$. If 3 does not divide t and $n \equiv 2 \pmod 4$, then each S_4–complement with core invariant Δ has a unique reduced invariant, and assigning to each such Frobenius complement its reduced invariant induces a bijection from the set of isomorphism classes of S_4–complements to the set of subgroups of $\mathbb{Z}_m^\bullet \times \mathbb{Z}_{n//6t}^\bullet$ of order dividing 2.*

8.3. COROLLARY. *Suppose $n \equiv 2 \pmod 4$ and $3 \nmid t$. Let s denote the number of distinct prime divisors of $mn//6t$. Then there are 2^s isomorphism classes of S_4–complements with core invariant Δ.*

PROOF. $\mathbb{Z}_m^\bullet \times \mathbb{Z}_{n//6t}^\bullet$ is a direct product of s cyclic groups of even order. Now apply the above theorem. □

As usual we proceed in a sequence of lemmas.

8.4. LEMMA. *Let G be an S_4–complement with core invariant Δ. Suppose we have a sequence u, v, w, x, y, z of elements of G such that $H := \langle u, v, x, y, z \rangle$ is an A_4–complement with the same core as G; that u, v, x, y, z is an r–sequence for H with invariant (b, h); that $|\langle w, v \rangle| = 16$; and that $w^2 = u$, $wzw^{-1} \in \langle x, y \rangle vz^2$, $wxw^{-1} = y^c x^g$, and $wyw^{-1} = y^a$ where a, $c \in \mathbb{Z}_m$ and $g \in \mathbb{Z}_n$. Then $3 \nmid t$, $b = 1$, $h = 1$, $g \equiv -1 \pmod{3_n}$, $g \equiv 1 \pmod{2t_n}$, $g^2 = 1$, $a^2 = 1$, $c(a+1) = 0$, $(m, a+1, a-1) = 1$, $m = (m, a+1)(m, a-1)$, $wvw^{-1} = uv$, and $wzw^{-1} = x^{-n//3}vz^2$. Moreover, there exists an integer a' with $a'(a-1) \equiv -c \pmod{(m, a+1)}$, and u, v, w, $y^{a'}x$, y, z is an r–sequence for G with invariant (a, g) for any such a'.*

PROOF. By Theorem 7.3, $n \equiv 2 \pmod 4$. Hence by Lemma 4.1A, $vwv^{-1} = w^{-1}$ and thus

$$wvw^{-1} = w(vw^{-1}v^{-1})v = w^2v = uv.$$

By Lemma 6.8 (applied with $s = 2$ to conjugation by w) we have $a^2 = 1$, $g^2 = 1$, $g \equiv 1 \pmod{t}$, $c(1+a) = 0$, $(m, a+1, a-1) = 1$, and $m = (m, a+1)(m, a-1)$. Lemma 6.9 then implies that $g \equiv 1 \pmod{t_n}$. Since $n \equiv 2 \pmod 4$ and $g^2 = 1$, then g is odd and $g \equiv 1 \pmod{2t_n}$.

By hypothesis we can write $wzw^{-1} = y^d x^e vz^2$ for some $d \in \mathbb{Z}_m$ and $e \in \mathbb{Z}_n$. Conjugating each term of the equation $zy = y^b z$ by w yields the relation
$$y^{d+ab^2 r^e} x^e vz^2 = y^d x^e vz^2 y^a = y^{ba+d} x^e vz^2$$
so $br^e = 1$. By Lemma 7.8 then, $r^{(n//3)+3e} = 1$, so $(n//3) + 3e \equiv 0 \pmod{t}$. Hence 3 does not divide t, as claimed. Thus t divides $n//3$, so $3e \equiv 0 \pmod t$ and hence $e \equiv 0 \pmod t$. But then $r^e = 1$, so $b = 1$. Next conjugate the relation $zx = x^h z$ by w to deduce that modulo the normal subgroup $\langle y \rangle$ we have
$$x^{e+gh^2} vz^2 \equiv x^e vz^2 x^g \equiv x^{gh+e} vz^2$$
so that $h = 1$.

Now conjugate $x^{n//3} = z^3$ by w; by Lemma 6.7 since t divides $n//3$ we have
$$x^{g(n//3)} = \left(y^d x^e vz^2\right)^3 = y^{3d} x^{3e} z^6 = y^{3d} x^{3e+2n//3}.$$
Thus $d = 0$ (Lemma 5.3(B1)) and
$$(8) \qquad 3e \equiv (g-2)(n//3) \pmod n.$$
Thus $e \equiv 0 \pmod{n//3}$. We further claim $g \equiv -1 \pmod{3_n}$. This is trivial if $3 \nmid n$; if 3 is a factor of n then the congruence (8) implies that $g \equiv 2 \pmod 3$, so $(3, g-1) = 1$. But since $g^2 = 1$, then $(g-1)(g+1) \equiv 0 \pmod{3_n}$, so $g+1 \equiv 0 \pmod{3_n}$, as claimed. The relation $w^2 = u$ implies that modulo $\langle y \rangle$,
$$\begin{aligned} v^{-1} z &\equiv u(zu^{-1}z^{-1})z \equiv w^2 zw^{-2} \equiv wx^e vz^2 w^{-1} \\ &\equiv x^{ge} uv(x^e vz^2)^2 \equiv x^{ge+2e+n//3} v^{-1} z, \end{aligned}$$
so $e(g+2) + n//3 \equiv 0 \pmod n$. But $g \equiv -1 \pmod{3_n}$, so $e \equiv -n//3 \pmod{3_n}$. Since $e \equiv 0 \pmod{n//3}$, we conclude that $e \equiv -n//3 \pmod n$. Thus $wzw^{-1} = x^{-n//3} vz^2$.

It remains to prove the last sentence of the lemma. Since $a-1$ and $(m, a+1)$ are relatively prime, there exists $a' \in \mathbb{Z}$ with $a'(a-1) \equiv -c \pmod{(m, a+1)}$. But then $a'(a-1) \equiv -c \pmod m$ since $m = (m, a-1)(m, a+1)$ and $c \equiv 0 \pmod{(m, a-1)}$ (recall that $c(1+a) = 0$). Now set $\check{x} = y^{a'} x$. Then $u, v, \check{x}, y$ is clearly an r–sequence for $\langle u, v, x, y \rangle$ with trivial invariant. Indeed, $u, v, \check{x}, y, z$ is an r–sequence for H with trivial invariant since t divides $n//3$ and hence $\check{x}^{n//3} = x^{n//3} = z^3$ (Lemma 6.7). To verify that $u, v, w, \check{x}, y, z$ is an r–sequence for G with invariant (a, g) it remains only to observe that
$$w \check{x} w^{-1} = y^{a'a+c} x^g = (y^{a'} x)^g = \check{x}^g$$
since $g \equiv 1 \pmod t$ and $a'(a-1) \equiv -c \pmod m$. □

8.5. LEMMA. *Suppose G is an S_4–complement with core invariant Δ. Then $n \equiv 2 \pmod 4$ and 3 does not divide t. G has an r–sequence, and both the invariant and reduced invariant of any r–sequence for G have orders dividing 2.*

PROOF. By hypothesis there is a surjective homomorphism $\Upsilon : G \longrightarrow S_4$ whose kernel is a $\mathbb{Z}$-group with invariant Δ. Then $H := \Upsilon^{-1}(A_4)$ is an A_4–complement with the same core as G. Hence by Theorem 7.3, $n \equiv 2 \pmod 4$ and H has an

r–sequence u, v, x, y, z, say with invariant (b, h). There is a Sylow 2–subgroup of G containing $\langle u, v \rangle$. It must be a generalized quaternion group and hence have an element w of order 8 whose square must be in $\langle u, v \rangle$; we may assume without loss of generality that $w^2 = u$ (if necessary replace w by an element of the form $z^i w z^{-i} u^j$ where $i, j \in \{0, 1, 2\}$).

$\Upsilon(z)$ has order 3 and $\Upsilon(w)$ has order 4. Hence without loss of generality we may assume Υ is chosen so that $\Upsilon(z)$ fixes 4 and $\Upsilon(w) = (1234)$. Thus $\Upsilon(u) = \Upsilon(w^2) = (13)(24)$. We claim $\Upsilon(z) = (132)$. For suppose this is not true. Then $\Upsilon(z) = (123)$. Thus $\Upsilon(wzw^{-1}) = (234) = \Upsilon(uz^2)$, so $wzw^{-1} = \gamma uz^2$ for some $\gamma \in \langle x, y \rangle$. Since $zvz^{-1} = u$ and u commutes with γ (Definition 7.2), we have (conjugating by w)
$$u = \gamma u z^2 u v z^{-2} u^{-1} \gamma^{-1} = \gamma u^{-1} \gamma^{-1} = u^{-1},$$
a contradiction. Thus $\Upsilon(z) = (132)$, as claimed. Therefore
$$\Upsilon(wzw^{-1}) = (243) = \Upsilon(z^2 u) = \Upsilon(vz^2).$$
Hence $wzw^{-1} \in \langle x, y \rangle vz^2$. Of course we can write $wyw^{-1} = y^a$ and $wxw^{-1} = y^c x^g$ for some $a, c \in \mathbb{Z}_m$ and $g \in \mathbb{Z}_n$. Then Lemma 8.4 applies to tell us that $3 \nmid t$; that G has an r–sequence with invariant (a, g); and that both the invariant and reduced invariant of any r–sequence have order at most 2. □

8.6. PROPOSITION. *Suppose G and G_1 are S_4-complements with core invariant Δ. Suppose (a, g) and (a_1, g_1) are the invariants of r-sequences u, v, w, x, y, z and u_1, v_1, w_1, x_1, y_1, z_1 of G and G_1, respectively. Then the following are equivalent:*
  (A) *$G \cong G_1$;*
  (B) *$\langle (a, g + n//6t\mathbb{Z}) \rangle = \langle (a_1, g_1 + n//6t\mathbb{Z}) \rangle$;*
  (C) *$(a, g) = (a_1, g_1)$;*
  (D) *There is an isomorphism $G \longrightarrow G_1$ taking u, v, w, x, y, z to u_1, v_1, w_1, x_1, y_1, z_1, respectively.*

Note that the implication (A) $\Longrightarrow$ (B) (applied in the case $G = G_1$) says that the reduced invariant of an S_4–complement is unique.

PROOF. Clearly (D) $\Longrightarrow$ (A). We now show (A) $\Longrightarrow$ (B) $\Longrightarrow$ (C) $\Longrightarrow$ (D).

(B) $\Longrightarrow$ (C). This is immediate from Lemma 8.4, since the element g of $\mathbb{Z}_n$ is completely determined by the congruences $g \equiv 1 \pmod{2t_n}$, $g \equiv -1 \pmod{3_n}$ and $g \equiv (g + n//6t\mathbb{Z}) \pmod{n//6t}$.

(C) $\Longrightarrow$ (D). The invariant (a, g) determines a presentation of G by the generators u, v, w, x, y, z and the relations described by Definition 8.1 and Lemma 8.4. If $(a, g) = (a_1, g_1)$, then G_1 has an equivalent presentation, and hence is isomorphic to G.

(A) $\Longrightarrow$ (B). We may suppose without loss of generality that $G = G_1$. Let u, v, w, x, y, z and u_1, v_1, w_1, x_1, y_1, z_1 be r–sequences for G with invariants (a, g) and (a_1, g_1), respectively. Now S_4 has a unique subgroup of index 2 and hence by hypothesis G has a unique subgroup of index 2 containing its core. Thus $\langle u, v, x, y, z \rangle = \langle u_1, v_1, x_1, y_1, z_1 \rangle$, and hence $w_1 \in G \backslash \langle u, v, x, y, z \rangle$. Thus we can write $w_1 = y^c x^f \gamma w$ for some $c \in \mathbb{Z}_m$, $f \in \mathbb{Z}_n$, and $\gamma \in \langle u, v, z \rangle$ (so γ commutes with x and y). Similarly we can write $y_1 = y^b$ and $x_1 = y^d x^e$ for $b, d \in \mathbb{Z}_m, e \in \mathbb{Z}_n$. Then
$$y_1^{a_1} = w_1 y_1 w_1^{-1} = y^c x^f \gamma w y^b w^{-1} \gamma^{-1} x^{-f} y^{-c} = y_1^{ar^f}.$$

Thus $a_1 = ar^f$, so $1 = r^{2f}$, so $r^f = 1$ (recall $t = |r|$ is odd). Thus $a = a_1$. Next, modulo the normal subgroup $\langle y \rangle$ we have

$$x_1^{g_1} \equiv w_1 x_1 w_1^{-1} \equiv x^f \gamma w x^e w^{-1} \gamma^{-1} x^{-f} \equiv x_1^g,$$

so $g = g_1$. Condition (B) follows immediately. □

The next theorem completes the proof of Theorem 8.2 by establishing the surjectivity part of the bijection in the theorem. The integral and rational truncated group rings of S_4-complements are also described.

8.7. THEOREM. *Suppose $n \equiv 2 \pmod{4}$ and 3 does not divide t. Let S be a subgroup of $\mathbb{Z}_m^\bullet \times \mathbb{Z}_{n//6t}^\bullet$ of order dividing 2.*

(A) There is an A_4-complement H with core invariant Δ and trivial invariant. There exists $(a,g) \in \mathbb{Z}_m^\bullet \times \mathbb{Z}_n^\bullet$ with $S = \langle (a, g + n/6t\mathbb{Z}) \rangle$, $g \equiv -1 \pmod{3_n}$, and $g \equiv 1 \pmod{(2t)_n}$. The natural map $H \longrightarrow \mathbb{Z}\langle H \rangle$ is injective; let u, v, x, y, z be an r-sequence for its image. Then there is an automorphism ρ of $\mathbb{Z}\langle H \rangle$ mapping u, v, x, y, z to $u, uv, x^g, y^a, x^{-n//3}vz^2$, respectively. For each $\gamma \in \mathbb{Z}\langle H \rangle$, $\rho^2(\gamma) = u\gamma u^{-1}$.

(B) Let $\mathcal{D} = (\mathbb{Z}\langle H \rangle, \rho, 2, u)$, $w = \widehat{\rho}$, and $G = \langle u, v, w, x, y, z \rangle$. Then G is an S_4-complement with core invariant Δ, and u, v, w, x, y, z is an r-sequence for G with invariant (a,g) and reduced invariant S. The inclusion map $G \longrightarrow \mathcal{D}$ induces an isomorphism $\mathbb{Z}\langle G \rangle \longrightarrow \mathcal{D}$. Finally $\mathbb{Z}\langle G \rangle$ is free of rank $8(n,3)\phi(mn)$ as a $\mathbb{Z}$-module.

(C) $\mathbb{Q}\langle G \rangle$ is isomorphic to $\mathbb{Q}\mathcal{D}$, which is simple with dimension $8(n,3)\phi(mn)$ over $\mathbb{Q}$. Let E denote the center of $\mathbb{Q}\langle H \rangle$. If $3 \nmid n$ and S is trivial, then $\mathbb{Q}\mathcal{D}$ has degree $2t$ and center $E[(1-u)w]$; otherwise $\mathbb{Q}\mathcal{D}$ has degree $4t$ and center E^ρ.

The proof of Theorem 8.7 will show that if 3 does not divide n and S is trivial, then $E[(1-u)w] \cong E[\sqrt{2}] \neq E$.

PROOF. Since 3 does not divide t, then the trivial group is in the set $\mathcal{S}$ of Notation 7.1 and hence is the reduced invariant of an A_4-complement H with core invariant Δ (by Theorem 7.3). The invariant of H is then also trivial (Lemma 7.8). The existence of (a, g) follows from the Chinese Remainder Theorem (again recall that 3 does not divide t). The natural map $H \longrightarrow \mathbb{Z}\langle H \rangle$ is injective by Lemma 2.3. Hence the image of H has an r-sequence u, v, x, y, z. Now consider the sequence $u, uv, x^g, y^a, x^{-n//3}vz^2$. First note that $g^2 = 1$ since $g^2 \equiv 1$ modulo $n//6t$, modulo 3_n and modulo $(2t)_n$. Since $g \equiv 1 \pmod{t}$, then u, v, x^g, y^a is easily checked to be an r-sequence with trivial invariant for the V_4-complement $\langle u, v, x, y \rangle = \langle u, uv, x^g, y^a \rangle$. Also

$$\left(x^{-n//3}vz^2\right)^3 = x^{(-n//3)3}v^2u^2x^{2(n//3)} = x^{-n//3} = \left(x^g\right)^{n//3}$$

since $g \equiv -1 \pmod{3_n}$. Since t divides $n//3$, $x^{-n//3}vz^2$ commutes with x and y. Next,

$$\left(x^{-n//3}vz^2\right)u\left(x^{-n//3}vz^2\right)^{-1} = x^{-n//3}vuv^{-1}x^{n//3} = v = (uv)u$$

and $\left(x^{-n//3}vz^2\right)(uv)\left(x^{-n//3}vz^2\right)^{-1} = u$. Thus $u, uv, x^g, y^a, x^{-n//3}vz^2$ is also an r-sequence with trivial invariant. Hence by Lemma 7.9 there is an automorphism of the image of H in $\mathbb{Z}\langle H \rangle$ taking u, v, x, y, z to $u, uv, x^g, y^a, x^{-n//3}vz^2$. Then this automorphism extends to the automorphism ρ of $\mathbb{Z}\langle H \rangle$. To verify the last part of (A) we must show $u\gamma = \rho^2(\gamma)u$ for $\gamma = u, v, x, y$ and z. The cases when

$\gamma = u, v, x, y$ are easily checked using the identities $g^2 = 1$ and $a^2 = 1$. Finally, since $g(n//3) \equiv -n//3 \pmod{n}$, then

$$\begin{aligned}\rho^2(z)u &= x^{-g(n//3)}uv\big(x^{-n//3}vz^2\big)\big(x^{-n//3}vz^{-2}z^3z\big)uz^{-1}z \\ &= uvvx^{-n//3}vux^{n//3}vuz = uz\,.\end{aligned}$$

This completes the proof of (A).

The assertions of part (A) guarantee that $\mathcal{D}$ is well–defined (cf. Notation 2.6B). Theorems 7.11 and 7.13 say that $\mathbb{Z}\langle H\rangle$ is free of rank $4(n,3)\phi(mn)$ as a $\mathbb{Z}$–module. Also $[G:H] = 2$ clearly divides $|H|$, and the Sylow 2–subgroup $\langle w,v\rangle$ of G is a generalized quaternion group. Thus we can apply the Cyclic Extension Lemma 2.8 to deduce that G is a Frobenius complement and the inclusion $G \longrightarrow \mathcal{D}$ induces an isomorphism from $\mathbb{Z}\langle G\rangle$ to $\mathcal{D}$, which by construction is free of rank $8(n,3)\phi(mn)$ as a $\mathbb{Z}$–module. Now $\langle x,y\rangle$ is by construction a normal $\mathbb{Z}$–subgroup of G. $G/\langle x,y\rangle$ is generated by the cosets of wz and of z since

$$w = (wz)z^{-1}, \quad u = w^2, \quad \text{and} \quad v = z^{-1}uz\,.$$

One checks that modulo $\langle x,y\rangle$

$$(wz)^2 \equiv 1, \quad z^3 \equiv 1, \quad \text{and} \quad ((wz)z)^4 \equiv 1\,.$$

Hence $G/\langle x,y\rangle$ is a homomorphic image of S_4 [**B**, p. 119]; since $G/\langle x,y\rangle$ has order $2[H:\langle x,y\rangle] = 24$, then $G/\langle x,y\rangle \cong S_4$. Thus G is an S_4–complement with core $\langle x,y\rangle$ and therefore core invariant Δ. By construction u, v, w, x, y, z is an r–sequence with invariant (a,g) and core invariant S. (The automorphism ρ is just conjugation by w.) This completes the proof of part (B).

We have isomorphisms

$$\mathbb{Q}\langle G\rangle \cong \mathbb{Q}\mathcal{D} \cong (\mathbb{Q}\langle H\rangle, \rho, 2, u)$$

(Notation 2.6); thus by Albert's Theorem 2.5 $\mathbb{Q}\mathcal{D}$ is simple with center E^ρ if ρ has order 2 on E. Now ρ^2 is conjugation by u, so its restriction to E is trivial. Hence ρ has order 2 on E if and only if it is nontrivial on E. Now by Theorems 6.15, 7.11 and 7.13, $E \supset K := \mathbb{Q}[x^t, y]^\sigma$ where σ is the automorphism of $\mathcal{D}$ given by conjugation by x. If 3 divides n, then since $g \equiv -1 \pmod{3}$ we have $x^t \in E$ and $\rho(x^t) = x^{tg} \neq x^t$. Now suppose 3 does not divide n and S is nontrivial. Then the restriction of ρ to $\mathbb{Q}[x^t, y]$ is an automorphism of order exactly 2 (this is obvious if $a \neq 1$; if $g + n//6t\mathbb{Z} \neq 1$ then $g \not\equiv 1 \pmod{n/t}$, so $\rho(x^t) = x^{tg} \neq x^t$). Hence this restriction does not fix $\mathbb{Q}[x^t, y]^\sigma$ (otherwise $\rho|\mathbb{Q}[x^t, y] \in \langle\sigma|\mathbb{Q}[x^t, y]\rangle$, which is impossible since the restrictions of ρ and σ have orders 2 and t, respectively, and t is odd). Thus ρ has order 2 on E. When this happens the degree of $\mathbb{Q}\langle G\rangle$ is clearly twice that of $\mathbb{Q}\langle H\rangle$, which is $2t$ for all A_4–complements with core invariant Δ and trivial invariant. Thus if either S is nontrivial or 3 divides n, then indeed $\mathbb{Q}\langle G\rangle$ is simple with degree $4t$ and center E^ρ. Now suppose S is trivial and 3 does not divide n. Then $E = \mathbb{Q}[x^t, y]^\sigma$ by Theorem 7.11, which also implies that $\mathbb{Q}[u, v, x, y, z] = \mathbb{Q}[u, v, x, y]$. (Remember that x and y are in $\mathbb{Z}\langle H\rangle$, not H.) Now let $\eta = (1-u)w$, so that $\eta^2 = (-2u)u = 2$ (recall that $w^2 = u$). By hypothesis η commutes with x and y; it clearly commutes with u, and also with v since by Lemma 8.4

$$v\eta v^{-1} = (v(1-u)v^{-1})v(wv^{-1}w^{-1})w = (1+u)v(uv)^{-1}w = \eta\,.$$

Hence $E[\eta]$ is contained in the centralizer of $\mathbb{Q}[u,v,x,y,z] = \mathbb{Q}\langle H\rangle$, which is simple with center E. Since $n \equiv 2 \pmod 4$, then 2 is unramified in $\mathbb{Q}[x^t, y]$ [**R**, 4B(1), p. 269] and hence in E. Thus $x^2 - 2$ is irreducible over E, so $E[\eta]$ is a quadratic field extension of E. Hence

$$[E[\eta] : E][\mathbb{Q}\langle H\rangle : E] = [\mathbb{Q}\mathcal{D} : E].$$

Further, since $(1+u)\eta = 2w$, then the set $E[\eta] \cup \mathbb{Q}\langle H\rangle$ generates $\mathbb{Q}\mathcal{D}$. Hence by [**P**, Proposition c, p. 165] $\mathbb{Q}\mathcal{D} \cong E[\eta] \otimes_E \mathbb{Q}\langle H\rangle$. This implies that $\mathbb{Q}\mathcal{D}$ is simple with center $E[\eta]$ as claimed. Thus the degree of $\mathbb{Q}\langle G\rangle$ is the same as that of $\mathbb{Q}\langle H\rangle$, namely, $2t$. In all cases the rational dimension of $\mathbb{Q}\langle G\rangle$ is the same as the rank of $\mathbb{Z}\langle G\rangle$, namely $8(n,3)\phi(mn)$. □

CHAPTER 9

Frobenius complements with core index 60

We continue to use the notation of Notation 6.2 and, when $n \equiv 2 \pmod 4$, the notation of Notation 7.10.

9.1. THEOREM. *There exists an A_5-complement with core invariant Δ if and only if $(mn, 60) = 2$. Suppose $(mn, 60) = 2$. The unique (up to isomorphism) A_5-complement with core invariant Δ is $J \times H_{120}$. Moreover, $\mathbb{Z}\langle J \times H_{120}\rangle$ is a free $\mathbb{Z}$-module of rank $8\phi(mn)$ and the natural map $J \times H_{120} \longrightarrow \mathbb{Z}\langle J\rangle \otimes \mathbb{Z}\langle H_{120}\rangle$ induces an isomorphism $\mathbb{Z}\langle J \times H_{120}\rangle \longrightarrow \mathbb{Z}\langle J\rangle \otimes \mathbb{Z}\langle H_{120}\rangle$. $\mathbb{Q}\langle J \times H_{120}\rangle$ is a simple algebra with dimension $8\phi(mn)$ over $\mathbb{Q}$ and with degree $2t$; its center is naturally isomorphic to $K[\sqrt{5}]$, a quadratic field extension of the center K of $\mathbb{Q}\langle J\rangle$.*

The rings $\mathbb{Z}\langle H_{120}\rangle$ and $\mathbb{Z}\langle J\rangle$ were analyzed in Theorems 3.2 and 5.2, respectively. (If $J = 1$ then $\mathbb{Z}\langle J\rangle = \mathbb{Z}$.) The proof of Theorem 9.1 will show that $J \times \langle -1\rangle$ is the core of G.

PROOF. First suppose $(mn, 60) = 2$. Let $G = J \times H_{120}$. $\mathbb{Z}\langle J\rangle$ is free of rank $\phi(mn)$ as a $\mathbb{Z}$-module (note that $\phi(1) = 1$; if $n \neq 2$ apply Theorem 5.2 to the triple $(m, n/2, T)$, noting that $\phi(mn) = \phi(mn/2)$). $\mathbb{Z}\langle H_{120}\rangle$ is isomorphic to $\mathbb{Z}[H_{120}]$ (Theorem 3.2), which as a $\mathbb{Z}$-module is free (cf. Lemma 3.5). Its rank is therefore the rational dimension of $\mathbb{Q}[H_{120}] = \mathbb{Q}[\mathbf{i},\mathbf{j},\sqrt{5}]$, which is 8. Since H_{120} and J have relatively prime orders we can apply the Direct Product Lemma 2.9 to deduce that G is a Frobenius complement and

$$\mathbb{Z}\langle G\rangle \cong \mathbb{Z}\langle J\rangle \otimes \mathbb{Z}\langle H_{120}\rangle,$$

so $\mathbb{Z}\langle G\rangle$ is free of rank $8\phi(mn)$ and $\mathbb{Q}\langle G\rangle$ has rational dimension $8\phi(mn)$. Next, $J \times \langle -1\rangle$ is a normal $\mathbb{Z}$-subgroup of G with

$$G/(J \times \langle -1\rangle) \cong H_{120}/\langle -1\rangle \cong A_5,$$

so $J \times \langle -1\rangle$ is the core of G and G is an A_5-complement. If $n = 2$ then G has core invariant $(1, 2, \langle 1\rangle) = \Delta$ (Lemma 5.3(B2)); suppose $n \neq 2$. Let $x = (x_0, -1)$ and $y = (y_0, 1)$. Then $\langle x, y\rangle$ is the core of G, $|x| = n$, $|y| = m$, and $xyx^{-1} = y^r$, so in this case also G has core invariant $(m, n, \langle r\rangle) = \Delta$.

$\mathbb{Q}[H_{120}] = \mathbb{Q}[\mathbf{i},\mathbf{j},\sqrt{5}]$ is a generalized quaternion algebra over $\mathbb{Q}[\sqrt{5}]$, so it is a simple algebra with center $\mathbb{Q}[\sqrt{5}]$. Next, $\mathbb{Q}\langle J\rangle$ is simple and its center, call it K, is isomorphic to a subfield of $\mathbb{Q}[\zeta_{mn/t}]$ (cf. Theorem 5.2 if $n \neq 2$). Since 5 does not divide mn, then $\sqrt{5} \notin K$ [**R**, 4B(1), p. 269]. Hence the center of $\mathbb{Q}\langle J\rangle \otimes \mathbb{Q}[H_{120}]$, namely $K \otimes \mathbb{Q}[\sqrt{5}]$, is a quadratic field extension of K isomorphic to $K[\sqrt{5}]$. Thus we have an isomorphism

$$\mathbb{Q}\langle G\rangle \cong \left(\mathbb{Q}[H_{120}] \otimes_{\mathbb{Q}[\sqrt{5}]} K[\sqrt{5}]\right) \otimes_{K[\sqrt{5}]} \left(\mathbb{Q}\langle J\rangle \otimes_K K[\sqrt{5}]\right)$$

(cf. [**P**, Exercise 3, p. 226]). Hence $\mathbb{Q}\langle G\rangle$ is simple with center isomorphic to $K[\sqrt{5}]$ [**P**, Proposition b((i) and (ii)), p. 226]. The degree of $\mathbb{Q}\langle G\rangle$ is the square root of

$$[\mathbb{Q}\langle G\rangle : \mathbb{Q}]/[K[\sqrt{5}] : \mathbb{Q}] = 8\phi(mn)/2[K : \mathbb{Q}] = 4[\mathbb{Q}\langle J\rangle : K] = 4t^2$$

(cf. Theorem 5.2), *i.e.*, the degree is $2t$.

Let us now suppose G is any A_5-complement with core invariant Δ; we must show that $(mn, 60) = 2$ and $G \cong J \times H_{120}$. Since A_5 is not solvable, then G is not solvable. Thus by a theorem of Zassenhaus [**Pa**, Theorem 18.6, p. 204], G has a normal subgroup M of index at most 2 which is isomorphic to $H \times H_{120}$ where H is a $\mathbb{Z}$-group of order relatively prime to 30. Now G has a subgroup which is an A_4-complement with the same core invariant Δ as G, so by Theorem 7.3 $n \equiv 2$ (mod 4). But then since $|G| = mn|A_5|$, $2_{|G|} = 8$. But 8 is already a divisor of the order of $H \times H_{120}$; hence $M = G$. Without loss of generality we may assume $G = H \times H_{120}$. We must show $H \cong J$. The core of G is $H \times \langle -1\rangle$ since this is a normal $\mathbb{Z}$-subgroup with factor group A_5 (cf. the Classification Theorem 1.4). Hence $H \times \langle -1\rangle$ has invariant Δ. If $n = 2$ then clearly $H = J = 1$ and we are done. Suppose $n \neq 2$. Suppose the $\mathbb{Z}$-group H has invariant $(m_1, n_1, \langle r_1\rangle)$ and r_1-sequence x_1, y_1. Let $x = (x_1, -1)$ and $y = (y_1, 1)$. Then $|x| = 2n_1$, $|y| = m_1$, $xyx^{-1} = y^{r_1}$ and $\langle x, y\rangle = H \times \langle -1\rangle$. Therefore $H \times \langle -1\rangle$ has invariant

$$(m, n, \langle r\rangle) = (m_1, 2n_1, \langle r_1\rangle),$$

so the invariant of H is

$$(m_1, n_1, \langle r_1\rangle) = (m, n/2, \langle r\rangle).$$

Thus H and J are isomorphic since they have the same invariant (Theorem 5.2). We also have that $(mn, 60) = 2$ since $(m_1n_1, 30) = 1$ and $n \equiv 2$ (mod 4). $\square$

CHAPTER 10

Frobenius complements with core index 120

As with other types of Frobenius complements, representations of S_5–complements by generators and relations will be developed at least implicitly (e.g., see Theorem 10.6B). However the emphasis will be put in this chapter on the close connection between S_5– and S_4–complements (cf. Proposition 10.7 below). The next definition exploits this connection to associate invariants with S_5–complements. We continue to use the notation of Notation 6.2 and, when $n \equiv 2 \pmod 4$, Notation 7.10.

10.1. DEFINITION. Suppose G is an S_5–complement with core invariant Δ. An element (a, g) of $\mathbb{Z}_m^\bullet \times \mathbb{Z}_n^\bullet$ is an *invariant* for G (and $\langle (a, g + n/\!/6t\mathbb{Z}) \rangle$ is a *reduced invariant* for G) if it is an invariant for a subgroup of G which is an S_4–complement with the same core as G.

10.2. THEOREM. *There is an S_5–complement with core invariant Δ if and only if $(mn, 60) = 2$. When $(mn, 60) = 2$, each S_5–complement with core invariant Δ has a unique reduced invariant and assigning to it its reduced invariant induces a bijection from the set of isomorphism classes of S_5–complements with core invariant Δ to the set of subgroups of $\mathbb{Z}_m^\bullet \times \mathbb{Z}_{n/\!/6t}^\bullet$ with order dividing 2.*

Before proving Theorem 10.2 we apply it to count isomorphism classes.

10.3. COROLLARY. *Suppose $(mn, 60) = 2$. The number of isomorphism classes of S_5–complements with core invariant Δ is 2^s, where s is the number of distinct prime factors of $mn/\!/6t$.*

We now prove a strong version of the assertion of Theorem 10.2 that each S_5–complement has a unique reduced invariant.

10.4. LEMMA. *Suppose $(mn, 60) = 2$. Let G be an S_5–complement with core invariant Δ. Then G has a subgroup H of index 5. Moreover every subgroup of G of index 5 is an S_4–complement with the same core as G and with the same core invariant, invariant, and reduced invariant as H.*

PROOF. The argument will depend on a simple property of S_5. (We thank Ed Bertram for a helpful discussion of subgroups of S_5 of index 5.)

Claim. Any subgroup S of S_5 of index 5 is conjugate to the subgroup S^* of S_5 of all permutations fixing 5.

Subproof. Since $|S| = 24$, S must contain a 3–cycle ξ and a Sylow 2–subgroup of S_5. Possibly replacing S by a conjugate we may assume without loss of generality that $S \supset \langle (1234), (13) \rangle$. If ξ has the form $(a\ b\ 5)$, then $\xi \delta \xi^{-1}$ has the form $(5\ c)$ where $\delta = (13)$ if $b \in \{1, 3\}$ and $\delta = (2, 4)$ if $b \in \{2, 4\}$. Thus the 5–cycle $(1234)(5\ c)$ is in S, contradicting that $[S_5 : S] = 5$. Hence ξ fixes 5 and so

$$S = \langle (1234), (13), \xi \rangle = S^*$$

(the middle group is contained in the others and has order at least $24 = |S| = |S^*|$). This completes the proof of the claim.

Since $S^* \cong S_4$ the Noether Isomorphism Theorem says G has a subgroup H of index 5 which is an S_4-complement with the same core N as G. Now let H_1 be any subgroup of G of index 5. For each odd prime p dividing mn let M_p be a Sylow p-subgroup of H_1. Then $M_p \subset N$ since every Sylow p-subgroup of G lies in the normal subgroup N for p an odd prime divisor of mn. But N is generated by all the M_p together with the unique element of G of order 2 (recall that $n \equiv 2 \pmod 4$ and m is odd), so $N \subset H_1$. Now the subgroups H/N and H_1/N of G/N ($\cong S_5$) are of index 5, so by the Claim above there exists $g \in G$ with $(gN)(H/N)(gN)^{-1} = H_1/N$. Thus $gHg^{-1} = H_1$. Since H and H_1 are conjugate, H_1 is an S_4-complement with the same core, core invariant, invariant and reduced invariant as H (Proposition 8.6). □

Part of the next lemma says that the hypothesis $(mn, 60) = 2$ of the previous lemma was redundant.

10.5. LEMMA. *Let G be an S_5-complement with core invariant Δ. Then $(mn, 60) = 2$. Moreover any S_5-complement with the same core invariant and reduced invariant as G is isomorphic to G.*

PROOF. Since A_5 is the unique subgroup of S_5 of index 2, then G has a unique subgroup A of index 2 which is an A_5-complement with the same core as G. Theorem 9.1 then implies that $(mn, 60) = 2$. Now let (a, g) be the invariant of G. By Proposition 8.6 it suffices to show that (a, g) (together with the core invariant Δ and core index 120) determine G up to isomorphism. By Theorem 9.1 we may assume without loss of generality that the subgroup A is an internal direct product of the groups H_{120} and J. By Lemma 4.2 there exists $w \in G$ with

(9) $\qquad w^2 = \mathbf{i},\ w\mathbf{j}w^{-1} = \mathbf{k},\ w\boldsymbol{\alpha}w^{-1} = (\boldsymbol{\alpha}\mathbf{i})^{-1},\ w\boldsymbol{\beta}w^{-1} = (\boldsymbol{\beta}\mathbf{i})^{-1}.$

The subgroup $JH_{24} = \langle \mathbf{i}, \mathbf{j}, \boldsymbol{\alpha}, x_0, y_0 \rangle$ has order $24(mn/2)$, so the group $S := \langle \mathbf{i}, \mathbf{j}, w, x_0, y_0, \boldsymbol{\alpha} \rangle$ has order $24mn$. Thus by Lemma 10.4 S is an S_4-complement with invariant (a, g) and the same core as G, namely $\langle -1x_0, y_0 \rangle$ (cf. the paragraph after the statement of Theorem 9.1). By Lemma 7.11 (last sentence) $\mathbf{i}, \mathbf{k}, -1x_0, y_0, \boldsymbol{\alpha}^2$ is an r-sequence for JH_{24}. By the last sentence of Lemma 8.4 we may assume without loss of generality that x_0 is chosen so that $\mathbf{i}, \mathbf{k}, w, -1x_0, y_0, \boldsymbol{\alpha}^2$ is an r-sequence for S with invariant (a, g). Then since g is odd, $wx_0w^{-1} = x_0^g$ and $wy_0w^{-1} = y_0^a$; these equations, together with the above list (9) of relations show exactly how G is determined as a cyclic extension of the direct product JH_{120}, and hence how is G determined up to isomorphism by its core invariant, core index, and invariant. □

In the above proof we showed that G was isomorphic to that extension of $J \times H_{120}$ by a cyclic group of order 2 associated with the element $(1, \mathbf{i})$ and automorphism (ψ_0, τ) of $J \times H_{120}$ where ψ_0 is the automorphism of Lemma 3.6A and τ is the automorphism of J taking x_0 to x_0^g and y_0 to y_0^a [**S**, 9.7.1, p. 250]. This is our only explicit reference to the machinery of cyclic extensions of groups; of course we have repeatedly made such constructions using the Cyclic Extension Lemma 2.8.

The next theorem will complete the proof of Theorem 10.2 and give the structure of the S_5-complements and their truncated group rings.

10.6. THEOREM. *Suppose that $(mn, 60) = 2$ and that S is a subgroup of $\mathbb{Z}_m^\bullet \times \mathbb{Z}_{n//6t}^\bullet$ of order dividing 2.*

(A) There exists $(a, g) \in \mathbb{Z}_m^\bullet \times \mathbb{Z}_n^\bullet$ with $S = \langle (a, g + n//6t\mathbb{Z}) \rangle$ and $g \equiv 1 \pmod{(2t)_n}$ and $g \equiv -1 \pmod{3_n}$. There also exists a unique automorphism ρ of $\mathbb{Z}\langle J \times H_{120} \rangle$ fixing $\overline{(1, \mathbf{i})}$ and mapping $\overline{(x_0, 1)}$, $\overline{(y_0, 1)}$, and $\overline{(1, \mathbf{j})}$ to $\overline{(x_0, 1)}^g$, $\overline{(y_0, 1)}^a$, and $\overline{(1, \mathbf{k})}$, respectively. Moreover ρ^2 acts as conjugation by $\overline{(1, \mathbf{i})}$.

(B) The subgroup $G := \langle \overline{(x_0, 1)}, \overline{(y_0, 1)}, \overline{(1, \mathbf{j})}, \overline{(1, \boldsymbol{\alpha})}, \overline{(1, \boldsymbol{\beta})}, \widehat{\rho} \rangle$ of $\mathcal{D}^\bullet$ where $\mathcal{D} := (\mathbb{Z}\langle J \times H_{120}\rangle, \rho, 2, \overline{(1, \mathbf{i})})$ is an S_5-complement with core invariant Δ, invariant (a, g), and reduced invariant S. The inclusion $G \longrightarrow \mathcal{D}$ induces an isomorphism $\theta : \mathbb{Z}\langle G \rangle \longrightarrow \mathcal{D}$. $\mathbb{Z}\langle G \rangle$ is free of rank $16\phi(mn)$ as a $\mathbb{Z}$-module.

(C) θ induces an isomorphism $\mathbb{Q}\langle G \rangle \longrightarrow \mathbb{Q}\mathcal{D}$. $\mathbb{Q}\mathcal{D}$ is simple with rational dimension $16\phi(mn)$ and degree $4t$. If E denotes the center of $\mathbb{Q}\langle J \times H_{120} \rangle$, then the center of $\mathbb{Q}\mathcal{D}$ is E^ρ.

The rings $\mathbb{Z}\langle J \times H_{120} \rangle$, $\mathbb{Q}\langle J \times H_{120} \rangle$, and E were analyzed in Theorem 9.1; Lemma 3.6 shows that ρ maps the square root of 5 in E to its additive inverse (recall $E \cong K[\sqrt{5}]$ where K is the center of $\mathbb{Q}\langle J \rangle$).

PROOF. The existence of g follows from the fact that $(2t)_n$, 3_n, and $n//6t$ are relatively prime. Since $g \equiv 1 \pmod{t}$, $x_0^g y_0^a x_0^{-g} = (y_0^a)^r$. Also $|x_0^g| = n/2$ and $|y_0^a| = m$. Hence there is a unique automorphism of $J = \langle x_0, y_0 \rangle$ taking x_0 to x_0^g and y_0 to y_0^a; by Lemma 3.6A there is a unique automorphism of H_{120} fixing $\mathbf{i}$ and mapping $\mathbf{j}$ to $\mathbf{k}$. Combining these automorphisms gives an automorphism ρ_0 of $J \times H_{120}$. Then ρ_0 induces the automorphism ρ of $\mathbb{Z}\langle J \times H_{120} \rangle$. Moreover ρ^2 is conjugation by $\overline{(1, \mathbf{i})}$ because ρ_0^2 is conjugation by $(1, \mathbf{i})$, as is easily checked (use Lemma 3.6A and the facts that $a^2 = 1$ and $g^2 = 1$). Thus the ring $\mathcal{D}$ is well-defined.

By the Cyclic Extension Lemma 2.8 (applied with "H" the image of $J \times H_{120}$ in $\mathbb{Z}\langle J \times H_{120} \rangle$) G is a Frobenius complement and the inclusion $G \longrightarrow \mathcal{D}$ induces an isomorphism $\mathbb{Z}\langle G \rangle \longrightarrow \mathcal{D}$. We now show G is an S_5-complement with invariant (a, g) and core invariant Δ. First, G contains an isomorphic copy of H_{120}, so it is not solvable. Further, it has an element of order 8, so it cannot be an A_5-complement (cf. Theorem 9.1). Thus it must be an S_5-complement by the Classification Theorem 1.4. The subgroup

$$H := \langle \overline{(x_0, 1)}, \overline{(y_0, 1)}, \overline{(1, \mathbf{i})}, \langle \overline{(1, \mathbf{j})}, \overline{(1, \boldsymbol{\alpha})}, \widehat{\rho} \rangle$$

has index 5 in G and hence by Lemma 10.4 is an S_4-complement with the same core, core invariant and invariant as G. Suppose the core of G has r-sequence x_1, y_1 and invariant $(m_1, n_1, \langle r_1 \rangle)$. Then $\langle x_1^2, y_1 \rangle$ is a normal subgroup of G of order $m_1 n_1/2$, which is relatively prime to $[G : \langle x_1^2, y_1 \rangle] = 2|S_5| = 240$. Hence it equals $\langle \overline{(x_0, 1)}, \overline{(y_0, 1)} \rangle$, since this supersolvable group also has order $mn/2 = |G|/240 = m_1 n_1/2$ [**S**, Exercise 9.3.20, p. 230]. Thus $\langle x_1, y_1 \rangle = \langle \overline{(x_0, -1)}, \overline{(y_0, 1)} \rangle$ (the right hand group is a subgroup properly containing $\langle \overline{(x_0, 1)}, \overline{(y_0, 1)} \rangle$, which has index 2 in $\langle x_1, y_1 \rangle$). Thus G has core invariant Δ. With Theorem 7.11 it is now routine to verify that

$$\overline{(1, \mathbf{i})}, \overline{(1, \mathbf{k})}, \widehat{\rho}, \overline{(x_0, -1)}, \overline{(y_0, 1)}, \overline{(1, \boldsymbol{\alpha}^2)}$$

is an r-sequence for H with invariant (a,g), so G does indeed have invariant (a,g). For example, to verify the last condition of Definition 8.1 we must check that

$$\widehat{\rho}(\overline{1,\boldsymbol{\alpha}^2})\widehat{\rho}^{-1} = (\overline{x_0,-1})^{-n/\!/3}(\overline{1,\mathbf{k}})(\overline{1,\boldsymbol{\alpha}^2})^2,$$

i.e., that $\rho_0((1,\boldsymbol{\alpha}^2)) = (1,\mathbf{k}\boldsymbol{\alpha})$ (since $3 \nmid n$). By Lemma 3.6A this is equivalent to saying $(\boldsymbol{\alpha}\mathbf{i})^{-2} = \mathbf{k}\boldsymbol{\alpha}$, which is easily verified using Lemma 3.4A.

From the construction of $\mathcal{D}$ we know that $\mathbb{Z}\langle G\rangle \cong \mathcal{D}$ is free of rank $16\phi(mn)$ as a $\mathbb{Z}$–module (recall from Theorem 9.1 that $\mathbb{Z}\langle J \times H_{120}\rangle$ is free of rank $8\phi(mn)$). It follows that $\mathbb{Q}\langle G\rangle$ has rational rank $16\phi(mn)$ and it is isomorphic to $\mathbb{Q}\mathcal{D} = \big(\mathbb{Q}\langle J \times H_{120}\rangle, \rho', 2, (\overline{1,\mathbf{i}})\big)$ where ρ' is the canonical extension of ρ to $\mathbb{Q}\langle J \times H_{120}\rangle$ (cf. Remark 2.6C). By Theorem 9.1 and Theorem 3.2 we have a natural isomorphism

$$\mathbb{Q}\langle J \times H_{120}\rangle \longrightarrow \mathbb{Q}\langle J\rangle \otimes \mathbb{Q}[\mathbf{i},\mathbf{j},\sqrt{5}\,].$$

By its construction ρ induces an automorphism of $\mathbb{Q}\langle J\rangle$ taking $\overline{x}_0$ to $\overline{x}_0^g$ and $\overline{y}_0$ to $\overline{y}_0^a$ and an automorphism of $\mathbb{Q}[\mathbf{i},\mathbf{j},\sqrt{5}\,]$ which by Lemma 3.6C maps $\sqrt{5}$ to $-\sqrt{5}$. The above isomorphism on $\mathbb{Q}\langle J \times H_{120}\rangle$ maps E to $K \otimes \mathbb{Q}[\sqrt{5}\,]$ where K is the center of $\mathbb{Q}\langle J\rangle$. Now the restriction of ρ to E has order exactly 2 (since $\rho(\sqrt{5}) = -\sqrt{5}$ and $a^2 = 1$ and $g^2 = 1$). Hence by Albert's Theorem 2.5 $\mathbb{Q}\mathcal{D}$ is simple with center E^ρ. The degree of $\mathbb{Q}\mathcal{D}$ is the square root of

$$[\mathbb{Q}\mathcal{D}:\mathbb{Q}]/[K[\sqrt{5}]^\rho:\mathbb{Q}] = 16\phi(mn)/[K:\mathbb{Q}] = 16[\mathbb{Q}\langle J\rangle:\mathbb{Q}]/[K:\mathbb{Q}] = 16t^2$$

(by Theorem 5.2). Thus the degree of $\mathbb{Q}\mathcal{D}$ is $4t$, as claimed. $\square$

The last result in this chapter highlights the close connection between isomorphism classes of S_4–complements and isomorphism classes of S_5–complements.

10.7. PROPOSITION. *There is a one-to-one correspondence between the set of isomorphism classes of S_5-complements and the set of isomorphism classes of S_4-complements whose cores have order relatively prime to 15. If G and H are an S_5-complement and an S_4-complement, respectively, then their isomorphism classes correspond if and only if H is isomorphic to a subgroup of G of index 5.*

PROOF. By Lemma 10.4 and the first part of Theorem 10.2, any S_5-complement G has a subgroup of index 5 which is an S_4-complement with the same core as G and hence with a core whose order is relatively prime to 15, and all such subgroups are isomorphic. Thus we have a well–defined function from the set of isomorphism classes of S_5-complements to the set of isomorphism classes of S_4-complements which have cores of order relatively prime to 15. Now let H be any such S_4-complement, say with invariant (a,g) and core invariant Δ. Since $(mn,60) = 2$ (note $n \equiv 2 \pmod 4$ by Theorem 8.2), there is an S_5-complement G with invariant (a,g) and core invariant Δ (Theorem 10.2). By Lemma 10.4 G has a subgroup of index 5 which is an S_4-complement with invariant (a,g) and core invariant Δ. Hence this subgroup is isomorphic to H. Therefore our map is surjective. If G_1 is another S_5-complement with a subgroup of index 5 isomorphic to H, then G_1 also has invariant (a,g) and core invariant Δ (Lemma 10.4). Thus $G \cong G_1$ (Theorem 10.2), so our map is also injective. $\square$

CHAPTER 11

Counting Frobenius Complements

The following table lists the number of isomorphism classes of Frobenius complements G subject to restrictions on the core index, call it I, of G and on the order $|G|$ of G.

	$|G| \leq 10^5$	$|G| \leq 5 \cdot 10^5$	$|G| \leq 10^6$
$I = 1$	256,349	1,494,276	3,191,063
$I = 4$	137,769	935,711	2,119,504
$I = 12$	2714	14,524	29,865
$I = 24$	3229	18,775	39,850
$I = 60$	223	1115	2230
$I = 120$	275	1599	3395
TOTAL	400,559	2,466,000	5,385,907

About 88.8% of the isomorphism classes of order at most 10^6 have even order (specifically, 4,782,903 out of 5,385,907). This suggests that the focus of this paper on Frobenius groups with abelian Frobenius kernel is not unduly restrictive.

The main ingredients in the construction of the above table were, first, the proposition below, which counts the number of proper Frobenius triples with certain given parameters, and, second, the "counting corollaries", by which name we refer collectively to Corollaries 6.6, 7.5, 8.3 and 10.3 which together with Theorems 5.2A and 9.1 allow one to easily compute the number of isomorphism classes of Frobenius complements with given core invariant in terms of these parameters. The calculations were organized around a nest of loops involving first picking a value for the order of the commutator subgroup of the core, then picking values of the exponents $e(i,j)$ of Part (B2) of the proposition below, and finally picking a value for the order of the commutator factor group of the core.

11.1. PROPOSITION. *Let $p(1), \ldots, p(s)$ be the prime divisors of an integer m such that $m \geq 3$.*

(A) Suppose $(m, n, \langle r \rangle)$ is a proper Frobenius triple. Set $t = |r|$. For each $i \leq s$ let $t(i) = |r + p(i)\mathbb{Z}|$. Then

(A1) $(m, n) = 1$, $n > 1$, and tt_0 divides n;

(A2) $t = \text{LCM}[t(1), \ldots, t(s)]$; and

(A3) if $i \leq s$, then $t(i)$ is a divisor of $(p(i) - 1)//m$ and $t(i) \neq 1$.

(B) Suppose now that n, t, and $t(1), \ldots, t(s)$ are any positive integers satisfying conditions (A1), (A2) and (A3).

(B1) Then the number of proper Frobenius triples of the form $(m, n, \langle r \rangle)$ with $t(i) = |r + p(i)\mathbb{Z}|$ for all $i \leq s$ is $\eta := \left(\prod_{i \leq s} \phi(t(i)) \right) / \phi(t)$.

(B2) Let $q(1),\ldots,q(s^*)$ denote the distinct prime factors of $\prod_{i\leq s}(p(i)-1)//m$ and for each $i \leq s$ write $t(i) = \prod_{j\leq s^*} q(j)^{e(i,j)}$. Then η equals

$$\prod_{j\leq s^*} q(j)^{L(j)-I(j)+\min\{1-M(j),0\}}(q(j)-1)^{I(j)-\min\{M(j),1\}}$$

where for each $j \leq s^*$, $L(j) = \sum_{i\leq s} e(i,j)$, $I(j) = \sum_{i\leq s} \min\{1,e(i,j)\}$, and $M(j) = \max\{e(1,j),\ldots,e(s,j)\}$.

In the statement of the above proposition $t(i) = |r+p(i)\mathbb{Z}|$ denotes the order of $r+p(i)\mathbb{Z}$ as an element of the multiplicative group $\mathbb{Z}^\bullet_{p(i)}$. Remark 11.2 below discusses some of the significance of these parameters.

PROOF. Condition (A1) follows directly from Definition 5.1(A and B). We can write $m = \prod_{i\leq s} p(i)^{e(i)}$; then we have a natural isomorphism

(10) $$\mathbb{Z}^\bullet_m \longrightarrow \prod_{i\leq s} \mathbb{Z}^\bullet_{p(i)^{e(i)}}.$$

Thus t is the least common multiple of the orders of $r(i) := r + p(i)^{e(i)}\mathbb{Z}$, $i \leq s$. Now the order of each $r(i)$ divides t and hence it is not divisible by $p(i)$. On the other hand $\mathbb{Z}^\bullet_{p(i)^{e(i)}}$ is cyclic of order $p(i)^{e(i)-1}(p(i)-1)$. Hence the order of $r(i)$ is $|r+p(i)\mathbb{Z}| = t(i)$ (the kernel of the canonical surjection $\mathbb{Z}^\bullet_{p(i)^{e(i)}} \longrightarrow \mathbb{Z}^\bullet_{p(i)}$ has order $p(i)^{e(i)-1}$). This proves (A2). Now fix for consideration any $i \leq s$. If $t(i) = 1$, then $r \equiv 1 \pmod{p(i)}$, contradicting that $(r-1,m) = 1$ (Lemma 5.3A). That $t(i)$ divides $p(i)-1$ and $(t(i),m) = 1$ follow from Lagrange's theorem and the fact that $(t,m) = 1$, respectively. This completes the proof of (A3).

Now suppose n, t and $(t(i) : i \leq s)$ satisfy conditions (A1), (A2), and (A3). Suppose $r \in \mathbb{Z}^\bullet_m$ has order relatively prime to m and that $|r+p(i)\mathbb{Z}| = t(i)$ for all $i \leq s$. We claim that $(m,n,\langle r\rangle)$ is a Frobenius triple. If some $p(i)$ divides $(m,r-1)$, then $1 = |r+p(i)\mathbb{Z}| = t(i)$, contradicting (A3). Hence $(m,r-1) = 1$. Since r has order relatively prime to m, the order of r equals the order of its image in $\prod_{i\leq s}\mathbb{Z}^\bullet_{p(i)}$, which is t. Since tt_0 divides n it follows that $r^{n/n_0} = 1$. That $(m,n,\langle r\rangle)$ is a Frobenius triple now follows from Lemma 5.3A.

The isomorphism (10) shows that the set of all $r \in \mathbb{Z}^\bullet_m$ satisfying the conditions of the previous paragraph is bijective with the set of s-tuples $(r_1,\ldots,r_s)$ where each r_i is an element of $\mathbb{Z}^\bullet_{p(i)^{e(i)}}$ of order $t(i)$. (Since $p(i) \nmid t(i)$, the order of any such r_i is the same as that of $r_i + p(i)\mathbb{Z}$ in $\mathbb{Z}^\bullet_{p(i)}$.) Since each $\mathbb{Z}^\bullet_{p(i)^{e(i)}}$ is cyclic, the number of such elements r is $\prod_{i\leq s}\phi(t(i))$. Each group generated by such an element is of course generated by $\phi(t)$ such elements. Hence the number of subgroups generated by such elements is $\left(\prod_{i\leq s}\phi(t(i))\right)/\phi(t)$, and hence this is the number of Frobenius triples $(m,n,\langle r\rangle)$ with $|r+p(i)\mathbb{Z}| = t(i)$ for all $i \leq s$. This completes the proof of (B1). The assertion (B2) follows from a routine computation using [NZ, Theorem 2.16, p. 48]. □

11.2. REMARK. Let Δ be the Frobenius triple $(m,n,\langle r\rangle)$ of Proposition 11.1A. Our focus above on the parameters m, n, and $t(i)$ ($i \leq s$) is only partly explained by the fact that Proposition 11.1B allows us to compute the number of Frobenius triples with these parameters. Just as important, these parameters determine the number of isomorphism classes of Frobenius complements with core invariant Δ. In fact by the counting corollaries the numbers of such isomorphism classes with given core index is determined by m, n, and t except in the case of core index 12 when 3 divides t. In this case the crucial condition (4) of Corollary 7.5 is satisfied if and only if for all $i \leq s$, either $3_{p(i)-1} = 1$ or $3_{p(i)-1}$ does not divide $t(i)$. (Note that $3_{t(i)}$ is the order of $r^{n//3} + p(i)\mathbb{Z}$.) The invariants $t(i)$ along with m and n are also enough to determine whether or not a 1–complement with invariant Δ is isomorphic to a subgroup of the group of units of a division ring (cf. Theorem 17.6).

CHAPTER 12

Maximal Orders

In this chapter we will prove Theorem 1.1 of the Introduction and give a representation of truncated group rings as crossed product algebras. The results in this chapter are crucial for the analysis of Frobenius groups with abelian kernel in the next three chapters.

12.1. NOTATION. For the remainder of this paper G will denote a Frobenius complement with core C and core invariant $\Delta = (m, n, T)$; as in Notation 6.2 we fix a generator r of T and set $t = |T|$. We also let $\mathcal{Z}$ and $\deg \mathbb{Q}\langle G \rangle$ denote the center and degree, respectively, of $\mathbb{Q}\langle G \rangle$.

12.2. THEOREM. *Suppose G is a Frobenius complement. Then $\mathbb{Q}\langle G \rangle$ is a simple ring with finite dimension as a $\mathbb{Q}$-algebra and with degree dividing $|G|$. Further, $\mathbb{Z}\langle G \rangle$ is nontrivial, finitely generated and free as a $\mathbb{Z}$-module; it is a $\mathbb{Z}$-order in $\mathbb{Q}\langle G \rangle$. Finally, $\mathbb{Z}_{(G)}\langle G \rangle$ is nontrivial, finitely generated and free as a $\mathbb{Z}_{(G)}$-module and is a $\mathbb{Z}_{(G)}$-order in $\mathbb{Q}\langle G \rangle$.*

The last sentence of Theorem 12.2 is an immediate consequence of the previous ones, which themselves were proved for each of the six types of Frobenius complements in Chapters 5 to 10. (Lemma 16.2 below gives a summary of some of the results of these chapters involving the rational dimension and the degree of $\mathbb{Q}\langle G \rangle$. It shows that the dimension is at most $|G|$ and the degree divides $[G:C]t$.)

The next corollary to Theorem 12.2 uses the fact that the construction of truncated group rings is functorial on the category of groups and injective homomorphisms.

12.3. COROLLARY. *Suppose H is a subgroup of a Frobenius complement G. Then the natural maps $G \longrightarrow \mathbb{Z}\langle G \rangle$, $\mathbb{Z}\langle H \rangle \longrightarrow \mathbb{Z}\langle G \rangle$, $\mathbb{Q}\langle H \rangle \longrightarrow \mathbb{Q}\langle G \rangle$, and $\mathbb{Z}\langle G \rangle \longrightarrow \mathbb{Q}\langle G \rangle$ are injective.*

PROOF. Theorem 12.2 says that $\mathbb{Z}\langle G \rangle$ is nontrivial and free as a $\mathbb{Z}$-module, so by Lemma 2.3 the map $G \longrightarrow \mathbb{Z}\langle G \rangle$ is injective. It also follows that $\mathbb{Z}\langle G \rangle \longrightarrow \mathbb{Q}\langle G \rangle$ is injective. The unitary homomorphism $\mathbb{Z}\langle H \rangle \longrightarrow \mathbb{Z}\langle G \rangle$ lifts to a unitary ring homomorphism $\mathbb{Q}\langle H \rangle \longrightarrow \mathbb{Q}\langle G \rangle$, which is injective since $\mathbb{Q}\langle H \rangle$ is simple and $\mathbb{Q}\langle G \rangle$ nontrivial. (Note that if H is trivial, then $\mathbb{Z}\langle H \rangle \cong \mathbb{Z}$.) Hence $\mathbb{Z}\langle H \rangle \longrightarrow \mathbb{Z}\langle G \rangle$ is also injective. □

It is perhaps worth recording a second corollary to Theorem 12.2 and the Direct Product Lemma 2.9.

12.4. COROLLARY. *A direct product of two nontrivial groups is a Frobenius complement if and only if the two groups are Frobenius complements of relatively prime orders.*

12.5. REMARK AND NOTATION. Let $L = \mathbb{Q}\langle C'\mathcal{Z}(C)\rangle$. $C'\mathcal{Z}(C)$ is a normal subgroup of G since the center $\mathcal{Z}(C)$ and commutator subgroup C' are characteristic subgroups of the normal subgroup C. We let x, y denote an r–sequence for C; then $C'\mathcal{Z}(C) = \langle x^t, y\rangle = \langle x^t y\rangle$, and $L = \mathbb{Q}[\overline{x}^t, \overline{y}] \cong \mathbb{Q}[\zeta_{mn/t}]$ by Theorem 3.2. We will further assume that x and y have been chosen so that they are part of an r–sequence for G if G is solvable (so that it has an r–sequence). Finally we use Corollary 12.3 to identify H with a subset of $\mathbb{Q}\langle H\rangle$ and L with a subfield of $\mathbb{Q}\langle H\rangle$ for any Frobenius complement H with core C and, more generally, to identify $\mathbb{Q}\langle J\rangle$ with a subring of $\mathbb{Q}\langle H\rangle$ for any subgroup J of H.

The next theorem shows that the center of $\mathbb{Q}\langle G\rangle$ is an abelian extension of $\mathbb{Q}$. It also gives implicitly a representation of $\mathbb{Q}\langle G\rangle$ as a crossed product algebra [**P**, §14.1, pp. 251–252]. It will be applied in the proof of Theorem 1.1 and in Chapters 13, 14, 15, and 17. Recall that a subfield E of $\mathbb{Q}\langle G\rangle$ containing $\mathcal{Z}$ is called *strictly maximal* if $[E : \mathcal{Z}]$ (or, equivalently, $[\mathbb{Q}\langle G\rangle : E]$) equals $\deg \mathbb{Q}\langle G\rangle$ [**P**, p. 236].

12.6. THEOREM. *There is an element ζ of $\mathbb{Z}_{(G)}\langle G\rangle$ such that $\zeta^{|G|} = 1$ and $E := \mathbb{Q}[\zeta]$ is a strictly maximal subfield of $\mathbb{Q}\langle G\rangle$ containing L. Moreover ζ can be chosen so that for each $\eta \in \mathrm{Gal}(E/\mathcal{Z})$ there exists $v_\eta \in \mathbb{Z}_{(G)}\langle G\rangle^\bullet$ such that for all $\tau, \rho \in \mathrm{Gal}(E/\mathcal{Z})$, $v_\tau v_\rho v_{\tau\rho}^{-1} \in \langle \zeta\rangle$ and τ is the restriction to E of conjugation by v_τ.*

12.7. REMARK. The proof of Theorem 12.6 will show that ζ can be chosen of order $\mathfrak{f} := smn/t$ where $s \leq 6$ and s divides $[G : C]$. $\mathfrak{f}$ is of course a multiple of the conductor of $\mathcal{Z}$; the conductor can equal $\mathfrak{f}$ (as when G is H_{48}, or cyclic of order not congruent to 2 modulo 4, or binary dihedral of order divisible by 8) or half $\mathfrak{f}$ (as when G is either H_{120} or cyclic or binary dihedral of other orders) or neither of these values (as when G is H_{24} or the 1–complement with invariant $(3, 4, \langle 2+3\mathbb{Z}\rangle)$).

In the construction below of the v_η of Theorem 12.6 we will always have $v_1 = 1$ (so the associated factor set is normalized).

We now prove the above theorem.

PROOF. Throughout the proof we let "$i \bmod j$" denote the remainder when i is divided by j (for $i, j \in \mathbb{Z}$ with $j > 0$). The proof will be broken into eleven cases; the only consolation the author can offer the reader is the fact that no further such casework will be needed until Chapter 15.

Case 1: $[G : C] = 1$. We use Theorem 5.2 and its notation to identify $\mathbb{Q}\langle G\rangle$ with $\mathbb{Q}\mathcal{A}$. Then $\mathbb{Q}\langle G\rangle$ has degree t and center L^σ. Hence $L = \mathbb{Q}[\zeta_{mn/t}]$ is itself a strictly maximal subfield of $\mathbb{Q}\langle G\rangle$ and we set $E = L$ and $\zeta = \zeta_{mn/t}$. The automorphism σ on L is precisely conjugation by x; it clearly generates $\mathrm{Gal}(E/\mathcal{Z})$, so we set $v_{\sigma^i} = x^i$ whenever $0 \leq i < t$. If also $0 \leq j < t$, then
$$v_{\sigma^i} v_{\sigma^j} v_{\sigma^i \sigma^j}^{-1} = x^i x^j x^{-((i+j) \bmod t)} \in \langle x^t\rangle$$
which is a subgroup of $\langle \zeta\rangle$.

The remaining cases are similar to Case 1, but with various complications.

Case 2: $[G : C] = 4$ and G has reduced invariant S with $|S| \leq 2$. In this case we use the notation and results of Theorem 6.14 and identify $\mathbb{Q}\langle G\rangle$ with $\mathbb{Q}\mathcal{B}$. The hypothesis on S implies that v commutes with x and y and hence with $\zeta_{mn/t}$. By Theorem 6.15 $\mathbb{Q}\langle G\rangle$ has center $(L^\sigma[v])^{\sigma_1}$ and degree $2t$, so that
$$E := L[v] = \mathbb{Q}[\zeta] \cong \mathbb{Q}[\zeta_{2mn/t}]$$

is a strictly maximal subfield of $\mathbb{Q}\langle G\rangle$, where $\zeta = v\zeta_{mn/t}$. Conjugation by x and by u induce automorphisms σ and σ_1 of E with orders t and 2, respectively. Both fix $\mathcal{Z}$ (recall that x and v commute) and hence they generate the Galois group $\mathrm{Gal}(E/\mathcal{Z})$. A typical element of the Galois group can be uniquely written in the form $\tau = \sigma^i \sigma_1^j$ (where $0 \le i < t$, $0 \le j < 2$) and can be obtained by conjugation by $v_\tau := x^i u^j$; each of the elements $v_\tau v_\rho v_{\tau\rho}^{-1}$ can be written in the form
$$\left(x^i u^j\right)\left(x^{i'} u^{j'}\right)\left(x^{(i+i')\bmod t} u^{(j+j')\bmod 2}\right)^{-1},$$
which is a power of $x^t = \zeta_{n/t} \in \langle\zeta\rangle$ since $u^2 = -1$ is a power of x^t, $uxu^{-1} = x^g$, and $g \equiv 1 \pmod{t}$.

Case 3: $[G:C] = |S| = 4$, where S is the reduced invariant of G. Again we use the notation of Theorem 6.14 and identify $\mathbb{Q}\langle G\rangle$ with $\mathbb{Q}\mathcal{B}$. By Theorem 6.15 $\mathbb{Q}\langle G\rangle$ has degree $4t$ and center $L^{\sigma,\sigma_0,\sigma_1}$, so in this case L is a strictly maximal subfield and we again set $E = L$ and $\zeta = \zeta_{mn/t}$. The Galois group of $E/\mathcal{Z}$ is generated by σ (conjugation by x), σ_0 (conjugation by v) and σ_1 (conjugation by u) since these homomorphisms of $\mathbb{Q}\langle G\rangle$ map E into itself, fix $\mathcal{Z}$, and generate a group of automorphisms of E of order $4t$. Each element of the Galois group can be uniquely written in the form $\tau = \sigma^i \sigma_0^j \sigma_1^k$ (where $0 \le i < t$, $0 \le j < 2$, $0 \le k < 2$) and is given by conjugation by $v_\tau := x^i v^j u^k$. Elements of $\mathbb{Q}\langle G\rangle$ of the form $v_\tau v_\rho (v_{\tau\rho})^{-1}$ can be written in the form
$$x^i v^j u^k \, x^{i'} v^{j'} u^{k'} \left(x^{(i+i')\bmod t} v^{(j+j')\bmod 2} u^{(k+k')\bmod 2}\right)^{-1},$$
which is a power of $\langle\zeta\rangle$ since uxu^{-1} and vxv^{-1} are congruent to x modulo $x^t = \zeta_{n/t}$, and both $u^2 = -1$ and $v^2 = x^{n//2}$ are in $\langle x^t\rangle \subset \langle\zeta\rangle$.

Case 4: $[G:C] = 12$ and $3 \nmid n$. We now use the notation and results of Theorem 7.11 and identify $\mathbb{Q}\langle G\rangle$ with
$$\mathbb{Q}[\mathbf{i},\mathbf{j}] \otimes \left(\mathbb{Q}[\zeta_{mn/t}], \sigma, t, \zeta_{n/t}^2\right).$$
Then $L = \mathbb{Q} \otimes \mathbb{Q}[\zeta_{mn/t}]$. The center of $\mathbb{Q}\langle G\rangle$ is $\mathbb{Q} \otimes \mathbb{Q}[\zeta_{mn/t}]^\sigma$ and the degree is $2t$. Let $\zeta = \mathbf{i} \otimes \zeta_{mn/t}^2$. Then
$$E := \mathbb{Q}[\mathbf{i}] \otimes \mathbb{Q}[\zeta_{mn/t}] = \mathbb{Q}[\zeta] \cong \mathbb{Q}[\zeta_{2mn/t}]$$
is a strictly maximal subfield. The Galois group of $E/\mathcal{Z}$ is generated by conjugation by $\mathbf{j} \otimes 1$ (call it ρ) and conjugation by $1 \otimes \widehat{\sigma}$ (call it δ). Elements of $\mathrm{Gal}(E/\mathcal{Z})$ can be uniquely written in the form $\tau = \delta^i \rho^k$ where $0 \le i < t$, $0 \le k < 2$; such τ is the restriction to E of conjugation by $v_\tau := (1 \otimes \widehat{\sigma})^i (\mathbf{j} \otimes 1)^k = \mathbf{j}^k \otimes \widehat{\sigma}^i$. For any τ, $\eta \in \mathrm{Gal}(E/\mathcal{Z})$ the element $v_\tau v_\eta v_{\tau\eta}^{-1}$ of $\mathbb{Q}\langle G\rangle$ can be written in the form
$$\left(\mathbf{j}^k \otimes \widehat{\sigma}^i\right)\left(\mathbf{j}^{k'} \otimes \widehat{\sigma}^{i'}\right)\left(\mathbf{j}^{(k+k')\bmod 2} \otimes \widehat{\sigma}^{(i+i')\bmod t}\right)^{-1}$$
which lies in
$$\pm\langle 1 \otimes \widehat{\sigma}^t\rangle \subset \langle \mathbf{i} \otimes \zeta_{mn/t}^2\rangle = \langle\zeta\rangle.$$

Case 5: $[G:C] = 12$, $3 \mid n$, and G has trivial reduced invariant. We now use the notation and results of Theorem 7.13, taking $S = 1$. We may identify $\mathbb{Q}\langle G\rangle$ with $\mathbb{Q}\mathcal{C}$. Let $A = (-1 + u + v + vu)/2$ and $\zeta = x^t yvAz$. As noted in the paragraph after the statement of Theorem 7.13, the ring $L[v, Az]$ is a cubic field extension of $L[v] = \mathbb{Q}[\zeta_{mn/t}, v] \cong \mathbb{Q}[\zeta_{2mn/t}]$ and Az is a root of unity of order $3(3_n)$. Since $\mathbb{Q}\langle G\rangle$ has degree $2t$ and center $L^\sigma[Az]$, it follows that
$$E := \mathbb{Q}[\zeta] = L[v, Az] \cong \mathbb{Q}[\zeta_{6mn/t}]$$

is a strictly maximal field extension of $\mathcal{Z}$. The automorphisms σ (conjugation by x) and σ_1 (conjugation by u) map E to itself, fix $\mathcal{Z}$, and have orders t and 2, respectively. Thus they generate $\text{Gal}(E/\mathcal{Z})$. Each member of the Galois group can be uniquely written in the form $\tau = \sigma^i \sigma_1^j$ ($0 \leq i < t$, $0 \leq j < 2$) and such τ acts by conjugation by $v_\tau := x^i u^j$. The elements of $\mathbb{Q}\langle G\rangle$ of the form $v_\tau v_\rho v_{\tau\rho}^{-1}$ are easily checked to all be in $\langle x^t \rangle \subset \langle \zeta \rangle$.

(The reader might note that the above case is one of the few in which the element ζ is not in the image of the map $G \longrightarrow \mathbb{Q}\langle G\rangle$.)

Case 6: $[G : C] = 12$, $3|n$, and G has nontrivial reduced invariant S. We again use the notation of Theorem 7.13 and identify $\mathbb{Q}\langle G\rangle$ with $\mathbb{Q}C$. The degree of $\mathbb{Q}\langle G\rangle$ is $6t$ and the center is $L^{\sigma,\tau}$, so

$$E := \mathbb{Q}[\zeta] = L[v] \cong \mathbb{Q}[\zeta_{2mn/t}]$$

is a strictly maximal subfield, where $\zeta = x^t y v$. We have automorphisms σ (conjugation by x), σ_1 (conjugation by u) and τ_1 (conjugation by Az, where A is defined as in the previous case) which map E into itself and fix $\mathcal{Z}$. (Note that conjugation by Az fixes v and acts on x^t and y exactly as τ does.) Since $(Az)^3 = z^3 = x^{n//3}$, then $\tau_1^3 = \sigma^{n//3}$, and so τ_1 has order $3(3_t)$. Thus $\langle \sigma, \sigma_1, \tau_1 \rangle$ has order $6t$, so it is the Galois group of $E/\mathcal{Z}$. Each element of the Galois group can be uniquely written in the form $\eta = \sigma^i \sigma_1^j \tau_1^k$ where $0 \leq i < t$, $0 \leq j < 2$, $0 \leq k < 3$; such η is precisely conjugation by $v_\eta := x^i u^j (Az)^k$. With $\rho = \sigma^{i'} \sigma_1^{j'} \tau_1^{k'}$ (where $0 \leq i' < t$, $0 \leq j' < 2$, $0 \leq k' < 3$) we have

$$\eta\rho = \sigma^{(i+i'+\delta(n//3)) \bmod t} \sigma_1^{(j+j') \bmod 2} \tau_1^{k+k'-3\delta}$$

where $\delta = 1$ if $k + k' \geq 3$ and $\delta = 0$ otherwise. Hence $v_\eta v_\rho v_{\eta\rho}^{-1}$ equals

$$x^i u^j (Az)^k x^{i'} u^{j'} (Az)^{k'} \left(x^{(i+i'+\delta(n//3)) \bmod t} u^{(j+j') \bmod 2} (Az)^{k+k'-3\delta} \right)^{-1}$$

which is a power of x^t and hence lies in $\langle \zeta \rangle$. (Recall that $zxz^{-1} = x^g$ where $g \equiv 1$ $(\bmod\ t)$; when $\delta = 1$ use the fact that $z^3 = x^{n//3}$.)

(The above case is one of the few in which some v_η is not in the image of the natural map $G \longrightarrow \mathbb{Q}\langle G\rangle$. In this case we used conjugation by Az instead of conjugation by z because the latter does not map E into itself.)

The next three cases deal with S_4-complements. In each of these cases we use the results and notation of Theorem 8.7 and we identify $\mathbb{Q}\langle G\rangle$ with $\mathbb{Q}\mathcal{D}$. Then L is identified with $\mathbb{Q}[x^t, y]$.

Case 7: $3 \nmid n$ and G is an S_4-complement with trivial reduced invariant. As usual let $\sigma \in \text{Aut}\, \mathbb{Q}\langle G\rangle$ be conjugation by x. Then $\mathbb{Q}\langle G\rangle$ has degree $2t$ and center $L^\sigma[\eta]$, where $\eta = (1 - u)w$. Let $\zeta = x^t y w$. Then

$$E := L[w] = \mathbb{Q}[\zeta] \cong \mathbb{Q}[\zeta_{4mn/t}]$$

is a strictly maximal subfield of $\mathbb{Q}\langle G\rangle$. The inner automorphisms σ and σ_0 (conjugation by v) of $\mathbb{Q}\langle G\rangle$ map E into itself, fix $\mathcal{Z}$, and have orders t and 2, respectively. (Note that $vwv^{-1} = w^{-1}$.) Thus they generate the Galois group of $E/\mathcal{Z}$. A typical element $\tau = \sigma^i \sigma_0^j$ (where $0 \leq i < t$, $0 \leq j < 2$) of $\text{Gal}(E/\mathcal{Z})$ is conjugation by $v_\tau := x^i v^j$; one easily checks that each $v_\tau v_\rho (v_{\tau\rho})^{-1}$ is in $\langle x^t \rangle \subset \langle \zeta \rangle$.

Case 8: $3 \nmid n$ and G is an S_4-complement with nontrivial reduced invariant. Then the degree of $\mathbb{Q}\langle G\rangle$ is $4t$ and the center is $L^{\rho,\sigma}$ where σ is conjugation by x

and ρ is conjugation by w. Let $\zeta = x^t yu$. Then
$$E := L[u] = \mathbb{Q}[\zeta] \cong \mathbb{Q}[\zeta_{2mn/t}]$$
is a strictly maximal subfield. The inner automorphisms σ_0 (conjugation by v), σ and ρ each map E into itself, fix $\mathcal{Z}$, and have orders 2, t and 2, respectively, on E. Thus they generate $\mathrm{Gal}(E/\mathcal{Z})$, a typical element of which can be written in the form $\tau = \sigma^i \sigma_0^j \rho^k$ where $0 \leq i < t$, $0 \leq j < 2$, $0 \leq k < 2$; τ is conjugation by $v_\tau := x^i v^j w^k$. Elements of $\mathbb{Q}\langle G \rangle$ of the form $v_\tau v_\eta v_{\tau\eta}^{-1}$ are all in $\langle u, x^t \rangle \subset \langle \zeta \rangle$. (Recall that $vwv^{-1} = w^{-1}$.)

Case 9: $3|n$ and $[G:C] = 24$. Then $\mathbb{Q}\langle G \rangle$ has degree $4t$ and center $(L^\sigma[Az])^\rho$ where, as before, $A = (-1+u+v+vu)/2$; σ is conjugation by x; and ρ is conjugation by w. Then
$$E := L[u, Az] = \mathbb{Q}[\zeta] \cong \mathbb{Q}[\zeta_{6mn/t}]$$
(where $\zeta = x^t yuAz$) is a strictly maximal subfield of $\mathbb{Q}\langle G \rangle$. The inner automorphisms σ_0 (conjugation by v), σ and ρ each map E to itself, fix $\mathcal{Z}$, and have orders 2, t and 2, respectively, on E. (Note that $wAzw^{-1} = x^{-n//3}(Az)^2$ and $x^{-n//3} \in \langle x^t \rangle$ since $3 \nmid t$.) Hence they generate the Galois group of $E/\mathcal{Z}$. As in the previous case, each $\tau = \sigma^i \sigma_0^j \rho^k$ (where $0 \leq i < t$, $0 \leq j < 2$, $0 \leq k < 2$) is conjugation by $v_\tau := x^i v^j w^k$ and elements of $\mathbb{Q}\langle G \rangle$ of the form $v_\tau v_\eta v_{\tau\eta}^{-1}$ are all in $\langle u, x^t \rangle \subset \langle \zeta \rangle$.

Case 10: $[G:C] = 60$. We use Theorem 9.1 and its notation, so we may identify $\mathbb{Q}\langle G \rangle$ with
$$\mathcal{E} := (\mathbb{Q}[\zeta_{mn/t}], \sigma, t, \zeta_{n/2t}) \otimes \mathbb{Q}[\sqrt{5}, \mathbf{i}, \mathbf{j}].$$
(We also use here the isomorphisms of Theorems 3.2 and 5.2D.) $\mathbb{Q}\langle G \rangle$ has degree $2t$ and center $\mathbb{Q}[\zeta_{mn/t}]^\sigma \otimes \mathbb{Q}[\sqrt{5}]$ and hence
$$E := \mathbb{Q}[\zeta_{mn/t}] \otimes \mathbb{Q}[\boldsymbol{\beta}] = \mathbb{Q}[\zeta] \cong \mathbb{Q}[\zeta_{5mn/t}]$$
is a strictly maximal subfield of $\mathbb{Q}\langle G \rangle$, where $\zeta = \zeta_{mn/t} \otimes \boldsymbol{\beta}$ ($\boldsymbol{\beta}$ is defined at the beginning of Chapter 3). The inner automorphisms σ (conjugation by $\hat{\sigma} \otimes 1$) and σ_0 (conjugation by $1 \otimes \mathbf{k}$) of $\mathbb{Q}\langle G \rangle$ map E into itself, fix $\mathcal{Z}$, and have orders t and 2, respectively, on E. (The key fact here is that $\mathbf{k}\boldsymbol{\beta}\mathbf{k}^{-1} = \boldsymbol{\beta}^* = \boldsymbol{\beta}^{-1}$ by Lemma 3.4B.) Hence σ and σ_0 generate $\mathrm{Gal}(E/\mathcal{Z})$. A typical element of the Galois group has the form $\tau = \sigma^i \sigma_0^j$ (where $0 \leq i < t$, $0 \leq j < 2$) and is given by conjugation by $v_\tau := \hat{\sigma}^i \otimes \mathbf{k}^j$. A routine computation shows all elements of $\mathbb{Q}\langle G \rangle$ of the form $v_\tau v_\eta v_{\tau\eta}^{-1}$ are in $\langle \zeta_{n/t} \otimes 1 \rangle \subset \langle \zeta \rangle$.

Case 11: $[G:C] = 120$. Let $\mathcal{E}$ be the ring introduced in Case 10 above. We may identify $\mathbb{Q}\langle G \rangle$ with $(\mathcal{E}, \rho, 2, 1 \otimes \mathbf{i})$ where $\rho_0 \in \mathrm{Aut}(\mathbb{Q}[\zeta_{mn/t}], \sigma, t, \zeta_{n/2t})$ maps ζ_m and $\hat{\sigma}$ to ζ_m^a and $\hat{\sigma}^g$, respectively; $\rho_1 \in \mathrm{Aut}\,\mathbb{Q}[\mathbf{i}, \mathbf{j}, \sqrt{5}]$ is the automorphism of Lemma 3.6C; and $\rho = \rho_0 \otimes \rho_1 \in \mathrm{Aut}\,\mathcal{E}$ (Theorem 10.6). The degree of $\mathbb{Q}\langle G \rangle$ is $4t$ and its center is $\mathcal{Z} = (\mathbb{Q}[\zeta_{mn/t}]^\sigma \otimes \mathbb{Q}[\sqrt{5}])^\rho$. Hence the field E of Case 10 is also a strictly maximal subfield for $\mathbb{Q}\langle G \rangle$ in this case (recall that $\rho_1(\sqrt{5}) = -\sqrt{5}$). A routine computation using Lemmas 3.4 (A and B) and 3.6 shows that

(11) $$(\mathbf{i}\alpha^{-1})\rho_1(\boldsymbol{\beta})(\mathbf{i}\alpha^{-1})^{-1} = \mathbf{i}\alpha^{-1}\mathbf{i}^{-1}\boldsymbol{\beta}^{-1}\alpha\mathbf{i}^{-1} = \boldsymbol{\beta}^2.$$

It follows that conjugation by $(1 \otimes \mathbf{i}\alpha^{-1})\hat{\rho}$ is an automorphism, call it ρ_3, of $\mathbb{Q}\langle G \rangle$ which maps E into itself, fixes $\mathcal{Z}$ and has order 4. Also conjugation by $\hat{\sigma} \otimes 1$ (call it σ) maps E to itself, fixes $\mathcal{Z}$, and has odd order t. Hence $\mathrm{Gal}(E/\mathcal{Z}) = \langle \sigma, \rho_3 \rangle$. Each element $\tau = \sigma^i \rho_3^j$ (where $0 \leq i < t$, $0 \leq j < 4$) of the Galois group is just

conjugation by $v_\tau := (\widehat{\sigma} \otimes 1)^i ((1 \otimes i\boldsymbol{\alpha}^{-1})\widehat{\rho})^j$. Let $\eta = \sigma^{i'} \rho_3^{j'}$ where $0 \leq i' < t$, $0 \leq j' < 4$. Then
$$v_{\tau\eta} = (\widehat{\sigma} \otimes 1)^{(i+i') \bmod t} ((1 \otimes i\boldsymbol{\alpha}^{-1})\widehat{\rho})^{(j+j') \bmod 4}.$$
Since $(1 \otimes i\boldsymbol{\alpha}^{-1})\widehat{\rho}(\widehat{\sigma} \otimes 1) = (\widehat{\sigma} \otimes 1)^g (1 \otimes i\boldsymbol{\alpha}^{-1})\widehat{\rho}$, then
$$((1 \otimes i\boldsymbol{\alpha}^{-1})\widehat{\rho})^j (\widehat{\sigma} \otimes 1)^{i'} ((1 \otimes i\boldsymbol{\alpha}^{-1})\widehat{\rho})^{j'} ((1 \otimes i\boldsymbol{\alpha}^{-1})\widehat{\rho})^{-((j+j') \bmod 4)}$$
$$= (\widehat{\sigma} \otimes 1)^{i' g^j} ((1 \otimes i\boldsymbol{\alpha}^{-1})\widehat{\rho})^{j+j' - ((j+j') \bmod 4)}.$$

Now by Lemma 3.6 and Lemma 3.4A,
$$((1 \otimes i\boldsymbol{\alpha}^{-1})\widehat{\rho})^4 = ((1 \otimes i\boldsymbol{\alpha}^{-1})(1 \otimes i\boldsymbol{\alpha}\mathbf{i})(1 \otimes \mathbf{i}))^2 = (1 \otimes -\mathbf{k})^2 = -1.$$
It follows easily that $v_\tau v_\eta v_{\tau\eta}^{-1} \in \pm \langle \widehat{\sigma}^t \otimes 1 \rangle \subset \langle \zeta \rangle$, which completes the proof of Case 11 and the theorem. □

The proof of the above theorem gives us a specific representation of $\mathbb{Q}\langle G \rangle$ as a crossed product algebra. We now prove a general result about maximal orders in crossed product algebras. The following notation will be used. Suppose E/K is a Galois field extension with Galois group H. Suppose $\Phi : H \times H \longrightarrow E$ is a normalized factor set (so $\Phi(1,1) = 1$). Then $(E/K, \Phi)$ denotes the central simple K-algebra which as a vector space over E has a basis $(u(\sigma) : \sigma \in H)$ such that
$$e_1 u(\sigma_1) e_2 u(\sigma_2) = e_1 \sigma_1(e_2) \Phi(\sigma_1, \sigma_2) u(\sigma_1 \sigma_2)$$
for all $e_1, e_2 \in E$ and $\sigma_1, \sigma_2 \in H$ [**P**, §14.1, p. 252]. Note that $u(1)$ is the identity since Φ is normalized.

12.8. THEOREM. *Suppose E/K is a Galois field extension with abelian Galois group H, where E is an algebraic number field. Let d be a nonzero integer with some power in the relative discriminant of $E/\mathbb{Q}$. Suppose $\Phi : H \times H \longrightarrow E$ is a normalized factor set with image in $(\mathrm{Int}_E \, \mathbb{Z}[1/d])^\bullet$. Then $\sum_{\sigma \in H} (\mathrm{Int}_E \, \mathbb{Z}[1/d]) u(\sigma)$ is a maximal $\mathbb{Z}[1/d]$-order in $(E/K, \Phi)$.*

PROOF. For any subfield F of E we will let $I(F) = \mathrm{Int}_F \, \mathbb{Z}[1/d]$, the integral closure of $\mathbb{Z}[1/d]$ in F. In particular we have $I(\mathbb{Q}) = \mathbb{Z}[1/d]$.

Let $A = (E/K, \Phi)$. Note that $\sum_{\sigma \in H} I(E) u(\sigma)$ is a subring of A and indeed an $I(\mathbb{Q})$-order. By [**Re**, Theorem 10.5 (iii), p. 128] it suffices to show that $\sum_{\sigma \in H} I(E) u(\sigma)$ is a maximal $I(K)$-order in A. Now suppose $\mathfrak{p}$ is any maximal ideal of $I(E)$. Then by [**Re**, Corollary 11.6, p. 134] it suffices to show that $I(K)_\mathfrak{p} \otimes_{I(K)} \sum_{\sigma \in H} I(E) u(\sigma)$ is a maximal $I(K)_\mathfrak{p}$-order in $K_\mathfrak{p} \otimes_K A$. Here and below for any subring D of E we let $D_\mathfrak{p}$ denote the completion of D in the $\mathfrak{p}$-adic topology.

Let H_0 denote the decomposition subgroup of H with respect to $\mathfrak{p}$ and let $F = E^{H_0}$ be the decomposition field. Let $C = (E/F, \Phi)$. (Strictly speaking Φ should be replaced in this definition of C by its restriction to $H_0 \times H_0$.) Let $f = [E : F]$ and $g = [F : K]$. Now no power of d is in $\mathfrak{p}$, so $\mathfrak{p}$ is unramified over $\mathbb{Q}$ [**R**, Theorem 1, p. 202] and hence it is unramified over K. Thus g is the number of extensions of $I(K) \cap \mathfrak{p}$ to a maximal ideal of $I(F)$ and f is the residue class degree of $\mathfrak{p}$ with respect to the extension E/K. Let $\xi_1, \ldots, \xi_g$ be a set of coset representatives in H for H/H_0; then
$$\xi_1(\mathfrak{p}) \cap I(F), \ldots, \xi_g(\mathfrak{p}) \cap I(F)$$

are the g distinct extensions of $\mathfrak{p} \cap I(K)$ to $I(F)$.

A is free with basis $u(\xi_1), \ldots, u(\xi_g)$ as a left C–module. (We identify C with the subring $\sum_{\sigma \in H_0} Eu(\sigma)$ of A.) Right multiplication by elements of A gives a map $\theta_0 : A \longrightarrow \text{End}_C A$ (so $\theta_0(a)(b) = ba$ for all $a, b \in A$). The basis $u(\xi_1), \ldots, u(\xi_g)$ has associated with it a map $\theta_1 : \text{End}_C A \longrightarrow M_g(C)$; for each $\gamma \in \text{End}_C A$, $\theta_1(\gamma)$ is the $g \times g$ matrix $[a_{ij}]$ where for all $i \leq g$, $\gamma(u(\xi_i)) = \sum_{j=1}^g a_{ij} u(\xi_j)$. Now let $C_\mathfrak{p} = (E_\mathfrak{p}/K_\mathfrak{p}, \Phi)$ (note that H_0 can be identified with the Galois group of $E_\mathfrak{p}$ over $F_\mathfrak{p} = K_\mathfrak{p}$). Then we have an inclusion map $\theta_2 : M_g(C) \longrightarrow M_g(C_\mathfrak{p})$. The composition $\theta_2 \theta_1 \theta_0 : A \longrightarrow M_g(C_\mathfrak{p})$ is a ring homomorphism. (This is a routine computation; let us briefly check multiplicativity. If $\theta_1 \theta_0(a) = [a_{ij}]$ and $\theta_1 \theta_0(b) = [b_{ij}]$, then for all $i \leq g$,

$$\theta_0(ab)(u(\xi_i)) = u(\xi_i) ab = \sum_k a_{ik} u(\xi_k) b$$

$$= \sum_k a_{ik} \sum_j b_{kj} u(\xi_j) = \sum_j \Big(\sum_k a_{ik} b_{kj}\Big) u(\xi_j)$$

so

$$\theta_1 \theta_0(ab) = \Big[\sum_k a_{ik} b_{kj}\Big] = [a_{ij}][b_{ij}] = \theta_1 \theta_0(a) \theta_1 \theta_0(b) \,.)$$

Let $\theta_3 : K_\mathfrak{p} \longrightarrow M_g(C_\mathfrak{p})$ be the natural homomorphism (map each a to $a[\delta_{ij}]$ where δ_{ij} denotes the usual Kronecker delta and $[\delta_{ij}]$ is therefore the $g \times g$ identity matrix). Composing $\theta_3 \otimes \theta_2 \theta_1 \theta_0$ with the multiplication map on $M_g(C_\mathfrak{p})$ yields a homomorphism

$$\theta : K_\mathfrak{p} \otimes_K A \longrightarrow M_g(C_\mathfrak{p})$$

which is injective since $K_\mathfrak{p} \otimes_K A$ is simple, and surjective since

$$[K_\mathfrak{p} \otimes_K A : K_\mathfrak{p}] = [A : K] = [E : K]^2$$
$$= f^2 g^2 = g^2 [C_\mathfrak{p} : F_\mathfrak{p}] = [M_g(C_\mathfrak{p}) : K_\mathfrak{p}] \,.$$

By [**Jz**, Theorem 1, p. 699] $\Lambda := \sum_{\sigma \in H_0} I(E)_\mathfrak{p} u(\sigma)$ is a maximal $I(K)_\mathfrak{p}$–order in $C_\mathfrak{p}$ and hence by [**Re**, Theorem 8.7, p. 110] $M_g(\Lambda)$ is a maximal $I(K)_\mathfrak{p}$–order in $M_g(C_\mathfrak{p})$. Therefore it suffices to show that

$$\mathcal{I} := \theta\Big(I(K)_\mathfrak{p} \otimes_{I(K)} \sum_{\sigma \in H} I(E) u(\sigma)\Big)$$

contains $M_g(\Lambda)$. This is clear if $g = 1$ (note that $I(E)_\mathfrak{p} = I(K)_\mathfrak{p} I(E)$), so suppose $g > 1$. We will prove $\mathcal{I} \supset M_g(\Lambda)$ by showing that for all $c \leq g$, $h \leq g$, $\sigma \in H_0$ and $e \in I(E)$, the matrix $eu(\sigma) E_{ch} = [eu(\sigma) \delta_{ic} \delta_{jh}]$ is in $\mathcal{I}$, where E_{ch} is the $g \times g$ matrix with a 1 in the (c, h) position and with all other entries 0. So pick such c, h, σ and e.

Suppose j and k are distinct positive integers less than or equal to g. Then ξ_j^{-1} and ξ_k^{-1} represent distinct cosets of H/H_0, so $\xi_j^{-1}(\mathfrak{p})$ and $\xi_k^{-1}(\mathfrak{p})$ restrict to distinct ideals of $I(F)$. Thus there exists $a \in I(F)$ with $a \in \xi_j^{-1}(\mathfrak{p}) \cap (1 + \xi_k^{-1}(\mathfrak{p}))$ (Chinese Remainder Theorem). Thus $\xi_j(a) - \xi_k(a) \in I(F) \setminus \mathfrak{p} \subset I(F)_\mathfrak{p}^\bullet = I(K)_\mathfrak{p}^\bullet$. (Since E/K is abelian, each ξ_i maps F into itself and hence maps $I(F)$ into itself.) Thus $\mathcal{I}$ contains the matrix

(12) $$\big(\xi_k(a) - \xi_j(a)\big)^{-1} \theta(1 \otimes a - \xi_j(a) \otimes 1) \,.$$

Note that $\theta(1 \otimes a)$ is the diagonal matrix whose i^{th} component is $\xi_i(a)$ (since $\theta_0(a)(u(\xi_i)) = u(\xi_i)a = \xi_i(a)u(\xi_i)$). Hence the matrix (12) is a diagonal matrix whose j^{th} term is
$$\bigl(\xi_k(a) - \xi_j(a)\bigr)^{-1}\bigl(\xi_j(a) - \xi_j(a)\bigr) = 0$$
and whose k^{th} term is
$$\bigl(\xi_k(a) - \xi_j(a)\bigr)^{-1}\bigl(\xi_k(a) - \xi_j(a)\bigr) = 1\,.$$
Taking the product of all these matrices for fixed k and for $j = 1, \ldots, g$, $j \neq k$, we see that E_{kk} is in $\mathcal{I}$ for all $k \leq g$.

For each $i \leq g$ we have a bijection s_i in the symmetric group S_g such that $\xi_i \xi_j = \tau_{ij} \xi_{s_i(j)}$ for all $j \leq g$, where $\tau_{ij} \in H_0$. Thus
$$\begin{aligned} u(\xi_i) \sum_{j \leq g} u(\xi_j) &= \sum_{j \leq g} \Phi(\xi_i, \xi_j) \Phi(\tau_{ij}, \xi_{s_i(j)})^{-1} u(\tau_{ij}) u(\xi_{s_i(j)}) \\ &= \sum_{j \leq g} e_j u(\tau_j) u(\xi_j) \end{aligned}$$
for some $e_j \in I(E)^\bullet$ and some $\tau_j \in H_0$. Therefore $\mathcal{I}$ contains a matrix, namely $\theta(1 \otimes \sum_{j \leq g} u(\xi_j))$, which for all $i \leq g$ and $j \leq g$ has (i,j)–entry of the form $bu(\rho)$ where $b \in I(E)^\bullet$ and $\rho \in H_0$. By multiplying such matrices on the right and left by those of the form E_{ii} we see $\mathcal{I}$ contains for all i and j a matrix of the form $bu(\tau)E_{ij}$ where $b \in I(E)^\bullet$ and $\tau \in H_0$. In particular for some such b and τ we have $bu(\tau)E_{ch} \in \mathcal{I}$. Note that since H is abelian, for any $a \in I(E)$, $\rho \in H_0$ and $i \leq g$,
$$u(\xi_i) a u(\rho) = \xi_i(a) \Phi(\xi_i, \rho) \Phi(\rho, \xi_i)^{-1} u(\rho) u(\xi_i),$$
so $\theta\bigl(1 \otimes au(\rho)\bigr)$ is a diagonal matrix with $\xi_i(a) \Phi(\xi_i, \rho) \Phi(\rho, \xi_i)^{-1} u(\rho)$ in the (i,i) position. Set $\rho = \sigma \tau^{-1}$ and
$$a = \xi_c^{-1}\bigl(e \Phi(\rho, \tau)^{-1} \rho(b)^{-1} \Phi(\rho, \xi_c) \Phi(\xi_c, \rho)^{-1}\bigr)\,,$$
so that $\theta\bigl(1 \otimes au(\rho)\bigr) bu(\tau) E_{ch}$ ($\in \mathcal{I}$) is a scalar multiple of E_{ch} by the scalar
$$\xi_c(a) \Phi(\xi_c, \rho) \Phi(\rho, \xi_c)^{-1} u(\rho) bu(\tau) = eu(\sigma)\,.$$
Thus $eu(\sigma) E_{ch} \in \mathcal{I}$, as was required to be shown. $\square$

Before proving Theorem 1.1 we set out the notation to be used in the proof, since this notation will be used again in the next three chapters.

12.9. REMARK AND NOTATION. We let $E = \mathbb{Q}[\zeta]$ be the maximal subfield of $\mathbb{Q}\langle G \rangle$ and $(v_\sigma : \sigma \in \mathcal{H})$ be the subset of $\mathbb{Q}\langle G \rangle$ constructed in the proof of Theorem 12.6, where $\mathcal{H} = \mathrm{Gal}(E/\mathcal{Z})$. For any subfield K of $\mathbb{Q}\langle G \rangle$, let I_K denote $\mathrm{Int}_K \mathbb{Z}_{(G)}$. Also let $\mathfrak{f}$ and s be as in Remark 12.7. Finally, for all $\sigma, \tau \in \mathcal{H}$, let $\Phi(\sigma, \tau) = v_\sigma v_\tau v_{\sigma\tau}^{-1}$. Then $(v_\sigma : \sigma \in \mathcal{H})$ is a basis for $\mathbb{Q}\langle G \rangle$ over E, Φ is a normalized factor set [**P**, Lemma, p. 251], and we have an isomorphism
$$\Upsilon : \mathbb{Q}\langle G \rangle \longrightarrow (E/\mathcal{Z}, \Phi)$$
carrying each sum $\sum_{\sigma \in \mathcal{H}} e_\sigma v_\sigma$ to $\sum_{\sigma \in \mathcal{H}} e_\sigma u(\sigma)$ (where $e_\sigma \in E$ for all $\sigma \in \mathcal{H}$).

We now give the proof of Theorem 1.1.

PROOF. Theorems 12.2 and 12.6 above include all of Theorem 1.1 except for the assertion that the $\mathbb{Z}_{(G)}$-order $\mathbb{Z}_{(G)}\langle G\rangle$ is maximal in $\mathbb{Q}\langle G\rangle$. The discriminant of E divides a power of $|\zeta|$ [**R**, 4B(2), p. 269]. Hence some power of $|G|$ is in the relative discriminant of $E/\mathbb{Q}$. Since ζ is integral over $\mathbb{Z}$, then Φ takes all of its values in $I_E^\bullet$. Therefore by the previous theorem, $\Lambda := \sum_{\sigma \in \mathcal{H}} I_E u(\sigma)$ is a maximal $\mathbb{Z}_{(G)}$-order in $(E/\mathcal{Z}, \Phi)$. By the choices of ζ and the v_σ ($\sigma \in \mathcal{H}$), the image of $\mathbb{Z}_{(G)}\langle G\rangle$ under Υ contains Λ (note that $I_E = \mathbb{Z}_{(G)}[\zeta]$). Since Λ is a maximal $\mathbb{Z}_{(G)}$-order, it follows that $\mathbb{Z}_{(G)}\langle G\rangle$ is a maximal $\mathbb{Z}_{(G)}$-order in $\mathbb{Q}\langle G\rangle$. □

The above proof of Theorem 1.1 yields the following corollary, which uses the notation of the preceding remark.

12.10. COROLLARY. *As an I_E-module $\mathbb{Z}_{(G)}\langle G\rangle$ has basis $(v_\sigma : \sigma \in \mathcal{H})$, and there is a unique I_E-algebra isomorphism from $\mathbb{Z}_{(G)}\langle G\rangle$ to $\sum_{\sigma \in \mathcal{H}} I_E u(\sigma)$ (a subring of $(E/\mathcal{Z}, \Phi)$) mapping v_σ to $u(\sigma)$ for each $\sigma \in \mathcal{H}$.*

We end this chapter with a fundamental lemma which will be used frequently (and often without citation) below. We use the notation of Remark 12.9.

12.11. LEMMA. *Let $\mathfrak{p}$ be a maximal ideal of $I_\mathbb{Q}$. Then $\mathfrak{p}$ is unramified in the field extension $E/\mathbb{Q}$ and the characteristic p of $I_\mathbb{Q}/\mathfrak{p}$ does not divide $[E : \mathcal{Z}]$.*

PROOF. Since by definition $|G|$ is a unit in $I_\mathbb{Q}$, therefore p does not divide $|G|$, and hence p divides neither $|\zeta|$ nor $\deg \mathbb{Q}\langle G\rangle = [E : \mathcal{Z}]$ (cf. Theorems 12.2 and 12.6). Thus $\mathfrak{p} = pI_\mathbb{Q}$ is unramified in $E/\mathbb{Q}$ [**R**, 4B(1), p. 269]. □

CHAPTER 13

Isomorphism Classes of Frobenius Groups with Abelian Frobenius Kernel

We will use notation of Notation 12.1 and Remark 12.9, so that in particular G denotes a Frobenius complement with core C and core invariant Δ. We will also write $R := \mathbb{Z}_{(G)}\langle G \rangle$. We study in this chapter Frobenius groups with Frobenius complement G and with abelian Frobenius kernel; it will be convenient to call such groups G–groups. Our object is to compute the set $Iso(G)$ of all isomorphism classes of G–groups in terms of the maximal ideals of the Dedekind domain $I_\mathcal{Z} = \text{Int}_\mathcal{Z}\,\mathbb{Z}_{(G)}$. To this end let $\mathcal{S}(G)$ denote the free abelian semigroup on the set P of nontrivial powers of maximal ideals of $I_\mathcal{Z}$. Let $\mathcal{S}(G)/\operatorname{Aut} G$ denote the set of orbits of elements of $\mathcal{S}(G)$ under the natural group action of $\operatorname{Aut} G$, the automorphism group of G. (Elements of $\operatorname{Aut} G$ induce automorphisms of $\mathbb{Q}\langle G \rangle$ which restrict to automorphisms of $I_\mathcal{Z}$ which, in turn, induce bijections of $\mathcal{S}(G)$.) Elements of $\mathcal{S}(G)$ will be written as formal sums $\eta = \sum_{i \leq \tau} n_i(\mathfrak{p}_i^{k_i})$ where $\tau > 0$ and for each $i \leq \tau$, $\mathfrak{p}_i$ is a maximal ideal of $I_\mathcal{Z}$, $n_i \geq 0$ (with some $n_j > 0$), and $k_i > 0$; every element of $\mathcal{S}(G)$ has such a representation with all the n_i equal to 1. The orbit in $\mathcal{S}(G)/\operatorname{Aut} G$ of $\eta \in \mathcal{S}(G)$ will be denoted $[\eta]$. We will also denote the isomorphism class of a group H by $[H]$.

The next lemma will make it easier to state our main theorem for this chapter.

13.1. LEMMA. [**BrH**, Theorem 8.2, p. 64]. *Suppose H is a G–group with Frobenius kernel M. Then M has a left R–module structure with $\bar{g}m = gmg^{-1}$ for all $g \in G$ and $m \in M$.*

PROOF. Since G acts on M (by conjugation) and M and G have relatively prime orders, then M has a $\mathbb{Z}_{(G)}G$–module structure with $\hat{g}m = gmg^{-1}$ for all $g \in G$ and $m \in M$. It therefore suffices to show that $\mathfrak{a}_G$ annihilates M (cf. Notation 2.1). If $1 \neq g \in G$ and $m \in M$, then

$$\hat{g}\big((\sum_{h \in \langle g \rangle} \hat{h})m\big) = \big(\sum_{h \in \langle g \rangle} \hat{h}\big)m$$

since $g\langle g \rangle = \langle g \rangle$. Hence $\big(\sum_{h \in \langle g \rangle} \hat{h}\big)m$ is indeed trivial (the action of G on M is without fixed points since H is a Frobenius group). That is, $\mathfrak{a}_G M$ is trivial. □

We now have the main result for this chapter.

13.2. THEOREM. *There is a bijection $\Psi : Iso(G) \longrightarrow \mathcal{S}(G)/\operatorname{Aut} G$ such that if H is a Frobenius group with Frobenius complement G and abelian Frobenius kernel M and if $\eta = \sum_{i \leq \tau} n_i(\mathfrak{p}_i^{k_i})$ is in $\mathcal{S}(G)$, then $\Psi([H]) = [\eta]$ if and only if for all $i \leq \tau$ there exists a maximal ideal $\mathfrak{b}_i$ of R and an indecomposable R–module M_i such that $\mathfrak{b}_i \cap I_\mathcal{Z} = \mathfrak{p}_i$, $\mathfrak{b}_i^{k_i}$ is the annihilator of M_i, and M is isomorphic as an R–module to the direct sum $M_1^{n_1} \oplus \cdots \oplus M_\tau^{n_\tau}$.*

A constructive definition of the map Ψ appears in Remark 13.5.

13.3. REMARK. Some language used informally in the Abstract and Introduction above will now be made precise. Suppose M and M^* are R–modules (so they are also G–modules). A G-*semi-linear isomorphism* from M to M^* is an ordered pair (f, λ) where $\lambda \in \operatorname{Aut} G$, $f : M \longrightarrow M^*$ is an additive bijection, and $f(gm) = \lambda(g)f(m)$ for all $g \in G$ and $m \in M$. Then the set of G–semi-linear isomorphism classes of finite R–modules is naturally bijective with the set of isomorphism classes of G–groups. (Associate with each such R–module M the semidirect product $M \rtimes G$.) While the point of view expressed in this remark, which is close to the spirit of [**BrH**], is not emphasized here, it is never far below the surface. The proof of the assertion of bijectivity above is mechanical, at least when one uses the fact, explained below in the proof of Theorem 13.2, that if two G–groups are isomorphic, then they admit an isomorphism which maps G to itself.

We now collect some of the ideal theory needed for the proof of the above theorem.

13.4. LEMMA. *(A) If $\mathfrak{a}$ is either a maximal ideal of R or the annihilator of a finite R–module, then $\mathfrak{a} \cap \mathbb{Z} \neq \{0\}$.*

(B) If $\mathfrak{a}$ is an ideal of R with $\mathfrak{a} \cap \mathbb{Z} \neq \{0\}$, then $R/\mathfrak{a}$ is a finite principal ideal ring, $\mathbb{Q}\mathfrak{a} = \mathbb{Q}\langle G\rangle$, and $\mathfrak{a}$ is finitely generated as a $\mathbb{Z}_{(G)}$–module.

(C) If $\mathfrak{a}$ is the annihilator of a finite indecomposable R–module, then $\mathfrak{a}$ is a nontrivial power of a maximal ideal of R.

We are using the term "ideal" in the usual two–sided ring–theoretic sense. Part (B) of the above lemma (together with the fact that R is a maximal order) implies that the ideal $\mathfrak{a}$ in (B) is a "two–sided R–ideal in $\mathbb{Q}\langle G\rangle$" in the sense of [**Re**, p. 193].

PROOF. Recall from Theorem 12.2 that R is finitely generated and free as a $\mathbb{Z}_{(G)}$–module (so any maximal ideal is nontrivial) and that $\mathbb{Q}\langle G\rangle$ is simple (so that for any nontrivial ideal $\mathfrak{a}$ of R, $Q\mathfrak{a} = \mathbb{Q}\langle G\rangle$). First suppose $\mathfrak{a}$ is an ideal of R with $\mathfrak{a} \cap \mathbb{Z} \neq \{0\}$, say with $0 < k \in \mathfrak{a} \cap \mathbb{Z}$. Since R is a maximal order (Theorem 1.1), it is a Dedekind prime ring [**MR**, Theorem 3.16, p. 148], and hence $R/\mathfrak{a}$ is a principal ideal ring [**MR**, Theorem 7.10(i), p. 164]. Without loss of generality the integer k may be chosen relatively prime to the unit $|G|$ of R, and hence R/Rk is a homomorphic image of the finite ring $\mathbb{Z}_k G$. Thus $R/\mathfrak{a}$ is finite, and also $\mathfrak{a}/Rk$ is finite. Since R, and hence Rk, are finitely generated as $\mathbb{Z}_{(G)}$–modules, therefore $\mathfrak{a}$ is finitely generated as a $\mathbb{Z}_{(G)}$–module. This completes the proof of (B).

Now suppose $\mathfrak{a}$ is a maximal ideal. Just suppose that $\mathfrak{a} \cap \mathbb{Z} = \{0\}$. If $0 \neq k \in \mathbb{Z}$, then $k + \mathfrak{a}$ must be a unit in $R/\mathfrak{a}$ since otherwise it generates a nontrivial proper ideal in a simple ring. Since $\mathbb{Q}\mathfrak{a} = \mathbb{Q}\langle G\rangle$, therefore for any $\gamma \in R$ we can write $n\gamma = m\delta$ for some nonzero $n, m \in \mathbb{Z}$ and $\delta \in \mathfrak{a}$. Then $\gamma + \mathfrak{a} = (n+\mathfrak{a})^{-1}(m\delta + \mathfrak{a}) = 0$. Thus $R = \mathfrak{a}$, contradicting the choice of $\mathfrak{a}$. Hence $\mathfrak{a} \cap \mathbb{Z} \neq \{0\}$.

Next suppose $\mathfrak{a}$ is the annihilator of a finite R–module M. By Lagrange's theorem $|M| \in \mathbb{Z} \cap \mathfrak{a} \neq \{0\}$. This completes the proof of (A). It remains to prove part (C), so suppose M is indecomposable. By [**Re**, Theorem 22.10, p. 193] (and parts (A) and (B) above), $\mathfrak{a}$ can be written as a product $\mathfrak{b}_1^{k_1} \cdots \mathfrak{b}_\tau^{k_\tau}$ of nontrivial powers of distinct maximal ideals $\mathfrak{b}_i$ of R, and this product is independent of the order of the factors. Suppose $\tau > 1$. Since maximal ideals are prime,

$\mathfrak{b}_1 + \mathfrak{b}_2 \cdots \mathfrak{b}_\tau = R$. Raising both sides of this equation to a sufficiently high power, we deduce that $R = \mathfrak{b}_1^{k_1} + \mathfrak{b}_2^{k_2} \cdots \mathfrak{b}_\tau^{k_\tau}$. M is an indecomposable $R/\mathfrak{a}$–module with trivial annihilator and $R/\mathfrak{a}$ is the direct sum of the ideals $\mathfrak{b}_1^{k_1}/\mathfrak{a}$ and $\mathfrak{b}_2^{k_2} \cdots \mathfrak{b}_\tau^{k_\tau}/\mathfrak{a}$. Hence M decomposes nontrivially as the direct sum of $\mathfrak{b}_1^{k_1} M$ and $\mathfrak{b}_2^{k_2} \cdots \mathfrak{b}_\tau^{k_\tau} M$, a contradiction. Hence $\tau = 1$, and so $\mathfrak{a} = \mathfrak{b}_1^{k_1}$. □

The next two remarks give constructions and notations which are central to the proof of Theorem 13.2. The first involves a constructive definition of the map Ψ.

13.5. REMARK. Let H be a G–group with Frobenius kernel M. By the Krull–Schmidt Theorem [**J1**, p. 115] (and Lemma 13.1) we can write M as a direct sum of indecomposable R–modules $M = \oplus_{i \leq \tau} M_i$ (uniquely up to isomorphism, except for the order of the summands). By Lemma 13.4C the annihilator of each M_i is a power $\mathfrak{b}_i^{k_i}$ of a maximal ideal $\mathfrak{b}_i$ of R. For each $i \leq \tau$, $\mathfrak{b}_i$ and k_i are uniquely determined by the annihilator $\mathfrak{b}_i^{k_i}$ [**Re**, Theorem 22.10, p. 193] and $\mathfrak{b}_i \cap I_Z$ is a maximal ideal of I_Z [**Re**, Theorems 10.5, p. 128, and 22.4, p. 191]. We write

$$\psi(H) = \sum_{i \leq \tau} \left((\mathfrak{b}_i \cap I_Z)^{k_i} \right) \in \mathcal{S}(G).$$

Note that $\Psi([H]) = [\psi(H)]$.

13.6. REMARK. Suppose M is a (left) R–module and $\sigma \in \operatorname{Aut} R$. Let σM denote the R–module with the same additive group as M and with scalar multiplication $\#_\sigma$ given by the formula $r \#_\sigma m = \sigma^{-1}(r) m$ for all $r \in R$, $m \in M$. One easily verifies that if $\mathfrak{a}$ is the annihilator of M, then $\sigma(\mathfrak{a})$ is the annihilator of σM. Also $\sigma^{-1}(\sigma M) = M$ and if N is a second R–module, then $\sigma(M \oplus N) = \sigma M \oplus \sigma N$. It follows that M is indecomposable if and only if σM is indecomposable. If $\tau \in \operatorname{Aut} G$ we will often let τ also denote the automorphism of R induced by τ and let τM denote the associated R–module.

13.7. LEMMA. *Suppose H and H^* are G–groups and $\sigma \in \operatorname{Aut} G$. Then there exists an isomorphism $H \longrightarrow H^*$ which restricts on G to σ if and only if $\psi(H^*) = \sigma \psi(H)$.*

PROOF. Suppose $\varphi : H \longrightarrow H^*$ is an isomorphism restricting on G to σ. We use the notation of Remark 13.5. For each $g \in G$ and $m \in M$,

$$\begin{aligned} \varphi(\overline{g} \#_\sigma m) &= \varphi(\sigma^{-1}(\overline{g}) m) = \varphi(\sigma^{-1}(g) m \sigma^{-1}(g^{-1})) \\ &= g \varphi(m) g^{-1} = \overline{g} \varphi(m) . \end{aligned}$$

Thus φ induces an R–module isomorphism $\sigma M \longrightarrow \varphi(M)$. Note that $\varphi(M)$ is the Frobenius kernel of H^*. By the previous remark, $\sigma M_1 \oplus \cdots \oplus \sigma M_\tau$ is the decomposition of σM into indecomposable R–modules, and hence up to isomorphism it is the decomposition of $\varphi(M)$ into indecomposable R–modules. For each $i \leq \tau$, $\sigma(\mathfrak{b}_i^{k_i}) = \sigma(\mathfrak{b}_i)^{k_i}$ is the annihilator of σM_i, and so

$$\psi(H^*) = \sum_{i \leq \tau} (\sigma \mathfrak{b}_i \cap I_Z)^{k_i} = \sigma \psi(H) .$$

Conversely, let us now suppose that $\psi(H^*) = \sigma \psi(H)$. We continue to use the notation of Remark 13.5 and similarly let $M^* = \oplus_{i \leq \tau^*} M_i^*$ denote the Frobenius kernel of H^* and its decomposition into indecomposable R–modules. Let $(\mathfrak{b}_i^*)^{k_i^*}$

denote the annihilator of M_i^* for each $i \leq \tau^*$. Our hypothesis then implies that $\tau = \tau^*$ and, after reindexing the M_i, that $h_i = k_i^*$ and $\sigma \mathfrak{b}_i \cap I_{\mathcal{Z}} = \mathfrak{b}_i^* \cap I_{\mathcal{Z}}$ for all $i \leq \tau^*$. But then $\sigma \mathfrak{b}_i = \mathfrak{b}_i^*$ for all $i \leq \tau$ [**Re**, Theorem 22.4, p. 191]. Hence σM_i and M_i^* are $R/(\mathfrak{b}_i^*)^{k_i}$–modules with the same trivial annihilator. Using Lemma 13.4 to apply [**J**, Theorem 44, p. 79] we deduce that σM_i and M_i^* are isomorphic as $R/(\mathfrak{b}_i^*)^{k_i}$–modules, and hence also as R–modules. Thus there is an R–module isomorphism $\mu : \sigma M \longrightarrow M^*$. Define $\varphi : H \longrightarrow H^*$ by setting $\varphi(mg) = \mu(m)\sigma(g)$ for all $m \in M$, $g \in G$. This map restricts to σ on G and it is a bijection since μ and σ are bijections. It is also the required isomorphism since for all $g, g' \in G$, m, $m' \in M$ we have

$$\begin{aligned}
\varphi(mgm'g') &= \varphi(mgm'g^{-1}gg') \\
&= \mu(m(\sigma(\bar{g})\#_\sigma m'))\sigma(gg') \\
&= \mu(m)\big(\sigma(\bar{g})\mu(m')\big)\sigma(g)\sigma(g') \\
&= \mu(m)\sigma(g)\mu(m')\sigma(g^{-1})\sigma(g)\sigma(g') \\
&= \varphi(mg)\varphi(m'g').
\end{aligned}$$

This completes the proof of the lemma. $\square$

The above lemma is related to Lemma 11.1 of [**BrH**, p. 72].

If H and H^* are G–groups, call them G–*isomorphic* if there is an isomorphism $H \longrightarrow H^*$ which fixes G. In the next proposition we compute the set $Iso(G)^*$ of G–isomorphism classes of G–groups. This is essentially the same thing as computing the set of isomorphism classes of finite R–modules. The proof will use the fact that if M is any R–module, then G acts on M (since G maps into $R^\bullet$) and hence we can form the semidirect product $M \rtimes G$.

13.8. PROPOSITION. ψ *induces a bijection from* $Iso(G)^*$ *to* $\mathcal{S}(G)$.

PROOF. Let H and H^* denote G–groups. One checks that an isomorphism $\varphi : H \longrightarrow H^*$ fixing G induces an R–module isomorphism of the Frobenius kernels of H and H^*. Hence $\psi(H) = \psi(H^*)$; this shows that ψ induces a well–defined map $Iso(G)^* \longrightarrow \mathcal{S}(G)$. Lemma 13.7 (applied with σ trivial) implies that this map is injective. We now prove surjectivity. Any element $\eta \in \mathcal{S}(G)$ can be written in the form $\sum_{i \leq \tau}(\mathfrak{p}_i^{k_i})$ where $1 \leq \tau \in \mathbb{Z}$ and for all $i \leq \tau$, $0 < k_i \in \mathbb{Z}$ and $\mathfrak{p}_i$ is a maximal ideal of $I_{\mathcal{Z}}$. Then there is a maximal ideal $\mathfrak{b}_1$ of R with $\mathfrak{b}_1 \cap I_{\mathcal{Z}} = \mathfrak{p}_1$ [**Re**, Theorem 22.4, p. 191]. Since $\mathfrak{b}_1 \cap \mathbb{Z} \neq \{0\}$, then $\mathfrak{b}_1^{k_1} \cap \mathbb{Z} \neq \{0\}$, so $R/\mathfrak{b}_1^{k_1}$ is a finite primary principal ideal ring; the radical is $\mathfrak{b}_1/\mathfrak{b}_1^{k_1}$ (Lemma 13.4). Thus $R/\mathfrak{b}_1^{k_1}$ can be identified with a matrix ring $M_u(B)$ where B is a completely primary ring [**J**, Theorem 31, p. 71]. Then $M_1 := (R/\mathfrak{b}_1^{k_1})E_{11}$ is an indecomposable $R/\mathfrak{b}_1^{k_1}$–module with trivial annihilator ideal, where E_{11} is the $u \times u$ matrix $[\delta_{i1}\delta_{j1}]$. (After all, $E_{11}M_1 \cong B$ is an indecomposable B–module; we use here the analysis of [**J**, pp. 78–79], adapted to left R–modules.) Then M_1 is an indecomposable R–module with annihilator $\mathfrak{b}_1^{k_1}$. Similarly we can find indecomposable R–modules $M_2, \ldots, M_\tau$ whose annihilators $\mathfrak{b}_2^{k_2}, \ldots, \mathfrak{b}_\tau^{k_\tau}$ satisfy $\mathfrak{b}_i \cap I_{\mathcal{Z}} = \mathfrak{p}_i$ whenever $2 \leq i \leq \tau$. Let $M = M_1 \oplus \cdots \oplus M_\tau$. G acts on M without fixed points. (This follows from an easy argument from [**BrH**, p. 65]: if $\bar{g}m = m$ where $1 \neq g \in G$, $m \in M$, then

$$\begin{aligned} m &= \bar{g}m = (1/|g|)(m + \bar{g}m + \cdots + \bar{g}^{|g|-1}m) \\ &= (1/|g|)\left(\sum_{h \in \langle g \rangle} \bar{h}\right)m = 0m = 0 \,.) \end{aligned}$$

Hence the semidirect product $M \rtimes G$ is a Frobenius group with Frobenius kernel $M \times 1$ and Frobenius complement $1 \times G$. Identifying $1 \times G$ with G we obtain a G–group H with

$$\psi(H) = \sum_{i \leq \tau} (\mathfrak{p}_i^{k_i}) = \eta\,.$$

This completes the proof of the proposition. □

We are now ready to give the proof of Theorem 13.2.

PROOF. First suppose $\varphi : H \longrightarrow H^*$ is an isomorphism of two G–groups. Then $\varphi(G)$ is a Frobenius complement of H^*, so there exists $b \in H^*$ with $b\varphi(G)b^{-1} = G$. Replacing φ by its composition with conjugation by b, we can assume without loss of generality that φ restricts on G to an automorphism σ of G. Lemma 13.7 (and Remark 13.5) then apply to say that $\psi(H^*) = \sigma\psi(H)$. Hence the map Ψ is well-defined; the same lemma applies to say it is injective. Finally Proposition 13.8 says that Ψ is surjective. □

The next theorem gives the group structures of the Frobenius kernels of G–groups in terms of their images under the bijection of Theorem 13.2. Knowledge of the orders of these groups will be needed in Chapter 15 when we compute the number of G–groups with specified order.

13.9. THEOREM. *Suppose $\eta = \sum_{i \leq \tau} (\mathfrak{p}_i^{k_i}) \in \mathcal{S}(G)$ where each $\mathfrak{p}_i$ is a maximal ideal of $I_\mathcal{Z}$. For each $i \leq \tau$ let p_i be the unique rational prime in $\mathfrak{p}_i$ and let f_i denote the ramification degree of p_i in the field extension $\mathcal{Z}/\mathbb{Q}$. Then the Frobenius kernel of any G–group J with $\Psi([J]) = [\eta]$ is isomorphic as a group to*

$$M(\eta) := \bigoplus_{i \leq \tau} (\mathbb{Z}_{p_i^{k_i}})^{f_i \deg \mathbb{Q}\langle G \rangle}\,.$$

The Frobenius kernel associated with η by Theorem 13.2 is the group M constructed in the proof of Proposition 13.8. Therefore in order to prove Theorem 13.9 it suffices to prove the following lemma which implies that M is the sum of the components $(\mathbb{Z}_{p_i^{k_i}})^{f_i \deg \mathbb{Q}\langle G \rangle}$.

13.10. LEMMA. *Let k be a positive integer and $\mathfrak{p}$ be a maximal ideal of $I_\mathcal{Z}$. Then*

(A) *$\mathfrak{p}R$ is a maximal ideal of R and $(\mathfrak{p}R) \cap I_\mathcal{Z} = \mathfrak{p}$;*
(B) *$I_\mathcal{Z}/\mathfrak{p}^k$ is isomorphic as an additive group to $\mathbb{Z}_{p^k}^f$ where p is the unique rational prime in $\mathfrak{p}$ and f is its residue class degree in $\mathcal{Z}/\mathbb{Q}$;*
(C) *$R/\mathfrak{p}^k R$ is isomorphic as a ring to $M_{\deg \mathbb{Q}\langle G \rangle}(I_\mathcal{Z}/\mathfrak{p}^k)$.*

PROOF. By Corollary 12.10 R is a free I_E–module of rank

$$\deg \mathbb{Q}\langle G \rangle = [\mathbb{Q}\langle G \rangle : E] = [E : \mathcal{Z}]$$

with basis $\{v_\sigma : \sigma \in \text{Gal}(E/\mathcal{Z})\}$. Also I_E is a free $I_\mathcal{Z}$-module with basis 1, ζ, $\zeta^2, \ldots, \zeta^{\deg \mathbb{Q}\langle G \rangle - 1}$; after all, this set is clearly linearly independent over $\mathcal{Z}$ and it spans I_E as an $I_\mathcal{Z}$-module since 1, ζ, $\zeta^2, \ldots, \zeta^{\mathfrak{f}-1}$ spans I_E as a $\mathbb{Z}_{(G)}$-module [**R**, 4B(3), p. 269] and the irreducible polynomial of ζ over $\mathcal{Z}$ has degree $\deg \mathbb{Q}\langle G \rangle$ and coefficients in $\mathcal{Z} \cap \mathbb{Z}_{(G)}[\zeta] = I_\mathcal{Z}$. Hence R is free of rank $(\deg \mathbb{Q}\langle G \rangle)^2$ as an $I_\mathcal{Z}$-module and has a basis containing 1. It follows that $(\mathfrak{p}^k R) \cap I_\mathcal{Z} = \mathfrak{p}^k$ and that $R/\mathfrak{p}^k R$ is free of rank $\deg \mathbb{Q}\langle G \rangle^2$ as an $I_\mathcal{Z}/\mathfrak{p}^k$-module.

In particular, $(\mathfrak{p}R) \cap I_\mathcal{Z} = \mathfrak{p}$. Hence we may regard $I_\mathcal{Z}/\mathfrak{p}$ as a subring of $R/\mathfrak{p}R$. Since $(v_\sigma : \sigma \in \text{Gal}(E/\mathcal{Z}))$ is a basis for R as an I_E-module, therefore $\mathfrak{p}I_E = (\mathfrak{p}R) \cap I_E$, so we can also regard $I_E/\mathfrak{p}I_E$ as a subring of $R/\mathfrak{p}R$. With this understanding we now turn to proving that $I_\mathcal{Z}/\mathfrak{p}$ is the center of $R/\mathfrak{p}R$.

Suppose ξ is in the center of $R/\mathfrak{p}R$. Write $\xi = \sum_{\sigma \in \text{Gal}(E/\mathcal{Z})} \xi_\sigma v_\sigma + \mathfrak{p}R$, where each ξ_σ is in I_E. Then by hypothesis $(\zeta + \mathfrak{p}R)\xi = \xi(\zeta + \mathfrak{p}R)$, so

$$\sum_{\sigma \in \text{Gal}(E/\mathcal{Z})} \zeta \xi_\sigma v_\sigma + \mathfrak{p}R = \sum_{\sigma \in \text{Gal}(E/\mathcal{Z})} \xi_\sigma \sigma(\zeta) v_\sigma + \mathfrak{p}R.$$

Since the v_σ form a basis for R as an I_E-module, then $\{v_\sigma + \mathfrak{p}R : \sigma \in \text{Gal}(E/\mathcal{Z})\}$ is a basis for $R/\mathfrak{p}R$ as an $I_E/\mathfrak{p}I_E$-module. Hence $\xi_\sigma(\zeta - \sigma(\zeta)) \in \mathfrak{p}R$ for all $\sigma \in \text{Gal}(E/\mathcal{Z})$. Since

$$1 + x + \cdots x^{\mathfrak{f}-1} = \prod_{1 \leq i < \mathfrak{f}} (x - \zeta^i),$$

therefore $\mathfrak{f}\zeta^{\mathfrak{f}-1} = \prod_{1 \leq i < \mathfrak{f}} \zeta - \zeta^{i+1}$. Since $\mathfrak{f}$ is a unit in R (it is a factor of $|G|$), therefore $\zeta - \zeta^j$ is a unit in R whenever $2 \leq j \leq \mathfrak{f}$. If $1 \neq \sigma \in \text{Gal}(E/\mathcal{Z})$, then $\sigma(\zeta) = \zeta^j$ where $2 \leq j < \mathfrak{f}$. Thus $\xi_\sigma \in \mathfrak{p}R$ when $\sigma \neq 1$. Hence we can write $\xi = \xi_0 + \mathfrak{p}R$ for some $\xi_0 \in I_E$.

Now let $\mathfrak{p}_1, \ldots, \mathfrak{p}_g$ denote the maximal ideals of I_E containing $\mathfrak{p}$, so that $\mathfrak{p}I_E = \mathfrak{p}_1 \cap \cdots \cap \mathfrak{p}_g$ (Lemma 12.11). For all $\sigma \in \text{Gal}(E/\mathcal{Z})$ and $j \leq g$, $(v_\sigma + \mathfrak{p}R)\xi = \xi(v_\sigma + \mathfrak{p}R)$, and hence

$$\xi_0 - \sigma(\xi_0) \in I_E \cap \mathfrak{p}R = \mathfrak{p}I_E \subset \mathfrak{p}_j,$$

so $\xi_0 + \mathfrak{p}_j = \sigma(\xi_0) + \mathfrak{p}_j$. It follows that $\xi_0 + \mathfrak{p}_j \in (I_\mathcal{Z} + \mathfrak{p}_j)/\mathfrak{p}_j \cong I_\mathcal{Z}/\mathfrak{p}$. (Note that every automorphism of $I_E/\mathfrak{p}_j$ fixing $(I_\mathcal{Z} + \mathfrak{p}_j)/\mathfrak{p}_j$ is induced by one in $\text{Gal}(E/\mathcal{Z})$.) Thus there exists $\gamma \in I_\mathcal{Z}$ with $\xi_0 + \mathfrak{p}_1 = \gamma + \mathfrak{p}_1$. Also for each $j \leq g$ there exists $\sigma \in \text{Gal}(E/\mathcal{Z})$ with $\sigma(\mathfrak{p}_1) = \mathfrak{p}_j$. Then

$$\xi_0 + \mathfrak{p}_j = \sigma(\xi_0) + \mathfrak{p}_j = \sigma(\xi_0 + \mathfrak{p}_1) = \sigma(\gamma + \mathfrak{p}_1) = \gamma + \mathfrak{p}_j,$$

so by Lemma 12.11

$$\xi_0 - \gamma \in \bigcap_{j \leq g} \mathfrak{p}_j = \mathfrak{p}I_E.$$

Hence $\xi = \xi_0 + \mathfrak{p}I_E = \gamma + \mathfrak{p}I_E \in I_\mathcal{Z}/\mathfrak{p}$. It follows that $I_\mathcal{Z}/\mathfrak{p}$ is indeed the center of $R/\mathfrak{p}R$.

Let $\mathfrak{q}$ be a maximal ideal of I_E containing $\mathfrak{p}$ and $E_\mathfrak{q}$ be a $\mathfrak{q}$-adic completion of E. We can pick a $\mathfrak{p}$-adic completion $\mathcal{Z}_\mathfrak{p}$ of $\mathcal{Z}$ contained in $E_\mathfrak{q}$, so $E_\mathfrak{q} = \mathcal{Z}_\mathfrak{p} E$. The ramification index of $E_\mathfrak{q}/\mathcal{Z}_\mathfrak{p}$ is trivial (apply Lemma 12.11). We may use the map Υ of Remark 12.9 to identify $\mathbb{Q}\langle G \rangle$ with $(E/\mathcal{Z}, \Phi)$, so the Schur index of $\mathcal{Z}_\mathfrak{p} \otimes_\mathcal{Z} \mathbb{Q}\langle G \rangle$ equals that of $(E_\mathfrak{q}/\mathcal{Z}_\mathfrak{p}, \Phi')$ (cf. [**Re**, Theorem 29.13, p. 248]; here Φ' is the restriction of the map Φ.) By [**Jz**, Theorem 1, p. 699], this Schur index divides the ramification index of $E_{\mathfrak{q}}/\mathcal{Z}_{\mathfrak{p}}$, so it is trivial. It follows from [**Re**, Theorem 22.14, p. 194] that $\mathfrak{p}R$ is a maximal ideal of R. This completes the proof of (A).

A routine induction on k shows that if $a_1, \ldots, a_f$ is a sequence of elements of $I_{\mathcal{Z}}$ representing a basis for $I_{\mathcal{Z}}/\mathfrak{p}$ over $\mathbb{Z}_p$, then it also represents a basis for $I_{\mathcal{Z}}/\mathfrak{p}^k$ as a $\mathbb{Z}_{p^k}$-module. The assertion (B) is an immediate consequence.

It remains to prove (C). As in the surjectivity part of the proof of Proposition 13.8 we have an isomorphism $\theta : R/\mathfrak{p}^k R \longrightarrow M_u(B)$ where B is a ring with nilpotent radical $\mathfrak{a}$ such that $B/\mathfrak{a}$ is a division ring. Since $B/\mathfrak{a}$ is finite, it is therefore a field. $M_u(\mathfrak{a})$ is an ideal of $M_u(B)$ with factor ring isomorphic to $M_u(B/\mathfrak{a})$, so $M_u(\mathfrak{a})$ is a maximal ideal of $M_u(B)$. But $\mathfrak{p}R/\mathfrak{p}^k R$ is the only maximal ideal of $R/\mathfrak{p}^k R$. Hence θ maps $\mathfrak{p}R/\mathfrak{p}^k R$ onto $M_u(\mathfrak{a})$, so $R/\mathfrak{p}R \cong M_u(B/\mathfrak{a})$ and

$$|B/\mathfrak{a}| = |\text{center of } M_u(B/\mathfrak{a})| = |\text{center of } R/\mathfrak{p}R| = |I_{\mathcal{Z}}/\mathfrak{p}|\,.$$

Therefore

$$\begin{aligned}(\deg \mathbb{Q}\langle G \rangle)^2 &= [R/\mathfrak{p}R : I_{\mathcal{Z}}/\mathfrak{p}] \\ &= [M_u(B/\mathfrak{a}) : B/\mathfrak{a}] = u^2\,,\end{aligned}$$

so $u = \deg \mathbb{Q}\langle G \rangle$. But

$$|B|^{u^2} = |M_u(B)| = |R/\mathfrak{p}^k R| = |I_{\mathcal{Z}}/\mathfrak{p}^k|^{(\deg \mathbb{Q}\langle G \rangle)^2}\,,$$

so $|B| = |I_{\mathcal{Z}}/\mathfrak{p}^k| = |I_{\mathcal{Z}} + \mathfrak{p}^k R/\mathfrak{p}^k R|$. But θ must map $I_{\mathcal{Z}} + \mathfrak{p}^k R/\mathfrak{p}^k R$ into the center of $M_u(B)$ and hence into the canonical image of B in $M_u(B)$. Thus $I_{\mathcal{Z}}/\mathfrak{p}^k \cong B$, and so $R/\mathfrak{p}^k R \cong M_{\deg \mathbb{Q}\langle G \rangle}(I_{\mathcal{Z}}/\mathfrak{p}^k)$. This completes the proof of (C). $\square$

We end this chapter with two applications, the first to nonsolvable Frobenius groups, and the second to Frobenius groups with nonabelian Frobenius kernel.

13.11. EXAMPLE. We apply the above theory to find all thirteen of the isomorphism classes of nonsolvable Frobenius groups of order at most 10^6. Methods of constructing these groups concretely are given in the next chapter. Since the kernel and the core of the complement of a Frobenius group are solvable, therefore a Frobenius group is nonsolvable if and only if its core index is 60 or 120 (since A_5 and S_5 are the only nonsolvable types, cf. Definition 6.1) and hence only if its Frobenius kernel is abelian. Let G be the Frobenius complement and M be the Frobenius kernel of such a Frobenius group $J = MG$. We will use the notation of this chapter associated with G. Let p denote any prime factor of $|M|$ and let f be the residue class degree of p in the field extension $\mathcal{Z}/\mathbb{Q}$. Note that $p \geq 7$ since p does not divide $|G| = mn[G:C]$. Also note that $m = 1$, since otherwise t is at least 7 and hence $\deg \mathbb{Q}\langle G \rangle$ is at least 14 (Theorems 9.1 and 10.6); this says $|M|$ is at least 7^{14} (Theorem 13.9), contradicting that $|J| \leq 10^6$.

Case 1: $[G:C] = 120$. Then 4 divides $\deg \mathbb{Q}\langle G \rangle$ (Theorem 10.6C), so by Theorem 13.9

$$10^6 \geq |J| \geq 2p^4[G:C] \geq 7^4 \cdot 240 = 576{,}240\,.$$

Since $2p^4[G:C]$ is a factor of $|J|$, it must therefore equal it; hence $p = 7$. Thus $|M| = 7^4$ and $|G| = 2[G:C]$ (so $n = 2$). Thus G is isomorphic to H_{240}, the unique S_5-complement with core of order 2 (cf. Theorem 10.6 and Example 4.3), and J is the unique G-group associated with the maximal ideal of $I_{\mathcal{Z}} = \mathbb{Z}[1/30]$ generated by 7 (cf. Theorem 13.2; note that $\mathcal{Z} = \mathbb{Q}[\sqrt{5}]^\rho = \mathbb{Q}$). That there actually exists such a group of order 576,240 also follows from the theorems cited above.

Case 2: $[G : C] = 60$ and $f > 1$. Then $\deg \mathbb{Q}\langle G\rangle \geq 2$ and $f \geq 2$, so by Theorem 13.9, $p^4 \leq |J|/|G| \leq 10^6/120 < 10^4$, so $p = 7$. Also
$$10^6 \geq |J| \geq 2[G : C](mn/2t)p^4 = 288{,}120(mn/2t).$$
Since any proper factor of $mn/2t$ must be at least 11 (Theorem 9.1), therefore $mn/2t = 1$, so $n = 2$. Hence $|M| = p^4$ and $2 = \deg \mathbb{Q}\langle G\rangle = f$. Thus $G \cong H_{120}$, the unique A_5–complement with two element core (Theorem 9.1) and J is the G–group associated with the unique maximal ideal of $I_\mathcal{Z} = \mathbb{Z}[1/30, \sqrt{5}\,]$ containing 7 (Theorem 13.2). That there actually is such a group of order 288,120 follows from the above cited theorems together with the fact that because $\binom{5}{7} = -1$, therefore the residue class degree of 7 in $\mathbb{Q}[\sqrt{5}\,]/\mathbb{Q}$ really is 2.

Case 3: $[G : C] = 60$, $f = 1$, and $n \neq 2$. Let q be an odd rational prime dividing n. Since $f = 1$, the residue class degrees of p in the extensions $\mathbb{Q}[\zeta_q]/\mathbb{Q}$ and $\mathbb{Q}[\sqrt{5}\,]/\mathbb{Q}$ must be trivial (note that $\mathcal{Z} \supset \mathbb{Q}[\zeta_q, \sqrt{5}\,]$ by Theorem 9.1). Thus $p \equiv 1 \pmod{q}$ (so $q < p$) and $1 = \binom{5}{p} = \binom{p}{5}$ (by the quadratic reciprocity theorem), so $p \equiv \pm 1 \pmod 5$. Just suppose $q \geq 11$. Then $10^6 \geq |J| \geq 120 \cdot 11 \cdot p^2$, so $p \leq 27$. This is impossible since no prime p congruent to ± 1 modulo 5 and less than 27 has $p - 1$ divisible by a prime larger than 7. Hence $q = 7$. Since $120qp^2 \leq 10^6$, then $p < 35$. Thus $p = 29$. Since $120p^2 q = 706{,}440$ divides $|J| \leq 10^6$, it must equal $|J|$ and hence $|G| = 120q$ and $|M| = 29^2$. Thus G must be the unique A_5–complement with core of order 14 (namely $H_{120} \times \langle \zeta_7\rangle$, cf. Theorem 9.1) and J is the unique G–group corresponding to the orbit in $\mathcal{S}(G)/\operatorname{Aut} G$ consisting of all the maximal ideals of $I_\mathcal{Z}$ containing the prime 29. (All these maximal ideals lie in the same orbit since the automorphisms of G induce all the automorphisms of $\mathcal{Z} = \mathbb{Q}[\sqrt{5}, \zeta_7]$. This can be proved directly, but it also follows immediately from Theorem 15.7 below.) That such a group J of order 706,440 exists follows from the fact that the prime 29 does indeed split in $\mathbb{Q}[\sqrt{5}, \zeta_7]$ and, of course, the theorems cited above.

We now have our last, easiest, and most prolific case.

Case 4: $[G : C] = 60$, $f = 1$ and $n = 2$. Then $\deg \mathbb{Q}\langle G\rangle = 2$, $\mathcal{Z} = \mathbb{Q}[\sqrt{5}\,]$ and, as in the previous case, since $f = 1$ we have $p \equiv \pm 1 \pmod 5$. Because $2[G : C]p^{\deg \mathbb{Q}\langle G\rangle} \leq |J| \leq 10^6$, therefore $p \leq 92$, so p is one of the primes

(13) $\qquad\qquad\qquad 11,\ 19,\ 29,\ 31,\ 41,\ 59,\ 61,\ 71,\ 79,\ 89.$

Since $|M|$ is a product of squares of the above numbers and $|M| \leq 10^6/2[G : C] \leq 8334$, then $|M|$ must actually be one of these squares. Hence $G \cong H_{120}$ and J must be one of the ten isomorphism classes of G–groups associated by Theorem 13.2 with the orbits of the maximal ideals of $I_\mathcal{Z} = \mathbb{Z}[1/30, \sqrt{5}\,]$ containing one of the ten primes in the list (13). (Note that each of these primes has two conjugate extensions to $I_\mathcal{Z}$, and these ten pairs of extensions constitute ten separate orbits of $\mathcal{S}(G)/\operatorname{Aut} G$.) The proof that all ten such groups of orders $120p^2$ exist (p on the list (13)) is routine.

The data in Example 15.3 below shows that the vast majority (that is, about 99.8%) of Frobenius groups of order at most 10^6 with abelian Frobenius kernel have cyclic Frobenius complement. The following proposition suggests that the same may be true for Frobenius groups of order at most 10^6 with nonabelian Frobenius kernel.

13.12. PROPOSITION. *Suppose J is a Frobenius group of order at most 10^9 with noncyclic Frobenius complement and nonabelian Frobenius kernel. Then J has Frobenius complement of order 171 and Frobenius kernel of order 7^6, or else Frobenius complement of order 63 and Frobenius kernel of order either 2^{12} or 2^{18}.*

The noncyclic Frobenius complement of order 63 is discussed briefly just after the statement of Theorem 17.6 below.

For the remainder of this chapter J will be as in the above proposition. G and K will denote a Frobenius complement and the Frobenius kernel for J, respectively. We use notation of this chapter in connection with G, so that G has core invariant $\Delta = (m, n, \langle r \rangle)$. The proof of the proposition will show at least implicitly that G is determined up to isomorphism by the order of J; the order of J also gives detailed information about the factors $K/K', K'/K'', \ldots$ of the derived series of K. For example if $|J| \leq 10^6$, then G is the unique noncyclic Frobenius complement of order 63 and $K/K' \cong K' \cong \mathbb{Z}_2^6$. The idea of the proof is that these factors are abelian Frobenius kernels for the Frobenius complement G, and hence their possible orders are determined by Theorem 13.9.

PROOF. Since K is nonabelian, G must have odd order and hence is a noncyclic $\mathbb{Z}$-group. Let p be the largest prime divisor of t. Since $p \nmid m$ and p is the order of an element of $\mathbb{Z}_m^\bullet$, there exists a prime divisor q of m with $q \equiv 1 \pmod{p}$, so that $q \geq 2p + 1$. For any prime divisor ρ of $|K|$ let $f(\rho)$ denote the residue class degree of ρ in the field extension $\mathcal{Z}/\mathbb{Q}$. We can write $\mathbb{Q}[\zeta_{mn/t}] \supset \mathcal{Z} \supset \mathbb{Q}[\zeta_{n/t}]$ (Theorem 5.2). Thus the residue class degree $|\rho + n/t\mathbb{Z}|$ of ρ in $\mathbb{Q}[\zeta_{n/t}]/\mathbb{Q}$ divides $f(\rho)$, so n/t divides $\rho^{f(\rho)} - 1$. Hence p divides $\rho^{f(\rho)} - 1$, and so $\rho^{f(\rho)} \geq p + 1 \geq 4$. Note similarly that $|\rho + m\mathbb{Z}|$ divides the residue class degree $|\rho + mn/t\mathbb{Z}|$ of ρ in $\mathbb{Q}[\zeta_{mn/t}]/\mathbb{Q}$, and hence divides $[\mathbb{Q}[\zeta_{mn/t}] : \mathcal{Z}]f(\rho) = tf(\rho)$. Thus m, and hence q, are divisors of $\rho^{tf(\rho)} - 1$.

Now let ρ be a prime such that K has a Sylow ρ-subgroup S which is not abelian (there must be such a Sylow subgroup since K is nonabelian and nilpotent [**T**, Theorem 1, p. 579]). Set $f = f(\rho)$. Since $S' \neq S''$, therefore by Theorem 13.9

$$|K| \geq |S/S'||S'/S''| \geq \rho^{2tf},$$

so

(14) $$|J| = mn|K| \geq p^2 q \rho^{2tf} \geq p^2(2p+1)\rho^{2pf}.$$

We use the inequality (14) repeatedly. Since $\rho^f \equiv 1 \pmod{p}$, if $p \geq 5$, then $\rho^f \geq 6$, so $10^9 \geq |J| \geq 5^2 \cdot 11 \cdot 6^{10}$, a contradiction. Thus $p = 3$. Hence by the choice of p, t is a power of 3. Recall that $\rho^f \geq 4$. Hence if 9 divides t, then $10^9 \geq |J| \geq 4^{18} > 10^9$. Therefore $t = 3$. Thus by the inequality (14), $10^9 \geq 9 \cdot 7 \cdot \rho^{6f}$, so $\rho^f \leq 13$. But $\rho^f \equiv 1 \pmod 3$ and if $\rho = 2$, then $2 = |\rho + 3\mathbb{Z}|$ divides f. Hence either $\rho = 2$ and $f = 2$; or $\rho = 7$ and $f = 1$; or $\rho = 13$ and $f = 1$.

First suppose that $\rho = 13$ and $f = 1$. As proved above, q is a divisor of $13^3 - 1$, whence $q = 61$ and $10^9 \geq 9 \cdot 61 \cdot 13^6$, a contradiction. Thus $\rho \neq 13$. Hence n/t divides either $2^2 - 1$ or $7^1 - 1$, whence $n/t = 3$. Thus $n = 9$ in all cases.

Now consider the case that $\rho = 7$ and $f = 1$. Since m divides $7^3 - 1$, therefore $m = 19$. Now suppose $|K|$ has a prime divisor ρ' other than ρ. As noted above $(\rho')^{f(\rho')} \geq 4$, so $10^9 \geq 9 \cdot 19 \cdot 7^6 \cdot 4^3$, a contradiction. Hence $|K| = 7^{3i}$ where $i \geq 2$. Since $10^9 \geq |J|$, we must have $i = 2$, so $|G| = mn = 171$ and $|K| = 7^6$.

13. ISOMORPHISM CLASSES OF FROBENIUS GROUPS

Finally suppose $\rho = 2$ and $f = 2$. Then m divides $\rho^{3f} - 1 = 63$, so $m = 7$ and $|G| = 63$. Suppose $|K|$ has a prime divisor ρ' other than ρ. Since $10^9 \geq |J| \geq 63 \cdot 4^6 \cdot (\rho')^{3f(\rho')}$, therefore $(\rho')^{f(\rho')} < 16$. However there is no such prime power which also satisfies the congruences $(\rho')^{f(\rho')} \equiv 1 \pmod{3}$ and $(\rho')^{3f(\rho')} \equiv 1 \pmod{7}$. Hence $|K| = 2^{6i}$ for some $i \geq 2$ with $10^9 \geq 63 \cdot 2^{6i}$. Hence i is 2 or 3, as was required to be shown. This completes the proof of the proposition. □

CHAPTER 14

Concrete Constructions of Frobenius Groups

We continue to use the notation set out in the first two sentences of the previous chapter, so that in particular G will denote a Frobenius complement and $R = \mathbb{Z}_{(G)}\langle G \rangle$. The construction of Frobenius groups with abelian Frobenius kernel in that chapter (cf. the proof of Proposition 13.8 and Lemma 13.10C) was not concrete in the sense that it depended on the choice, for each power $\mathfrak{a}$ of a maximal ideal of $I_{\mathcal{Z}}$, of an isomorphism $R/\mathfrak{a}R \longrightarrow M_{\deg \mathbb{Q}\langle G \rangle}(I_{\mathcal{Z}}/\mathfrak{a})$, but there was no indication of how to actually construct such isomorphisms. What is really needed is some explicit construction of an indecomposable R–module with annihilator $\mathfrak{a}R$ and this is precisely the purpose of the first lemma below. It yields in Theorem 14.2 a concrete way of representing every G–group. We also give here explicit constructions, which use little or nothing beyond elementary number theory, of all the Frobenius groups whose Frobenius complement is either H_{120} ($\cong SL(2,5)$) or a binary dihedral group. These constructions can be read independently of each other.

The next lemma shows for any power $\mathfrak{a}$ of a maximal ideal of $I_{\mathcal{Z}}$ how to put an R–module structure on $I_E/\mathfrak{a}I_E$ (or, more precisely, on an I_E–module $M(\mathfrak{a})$ which is naturally isomorphic to it), so that it becomes an indecomposable R–module with annihilator $\mathfrak{a}R$. Recall that the elements v_τ ($\tau \in \mathrm{Gal}(E/\mathcal{Z})$) form a basis for R as an I_E–module (cf. Corollary 12.10).

14.1. LEMMA. *Let $\mathfrak{p}$ be a maximal ideal of $I_{\mathcal{Z}}$ and let $\mathfrak{q}$ be a power of a maximal ideal of I_E with $\mathfrak{q} \cap I_{\mathcal{Z}} = \mathfrak{p}^k$ where $0 < k \in \mathbb{Z}$. Let E_d denote the decomposition field for $\mathfrak{p}$ in the field extension $E/\mathcal{Z}$.*

(A) $\mathrm{Gal}(E/\mathcal{Z})$ is the internal direct sum of a subgroup $\mathcal{G}$ which is disjoint from $\mathrm{Gal}(E/E_d)$ and a cyclic group $\langle \tau \rangle$ containing $\mathrm{Gal}(E/E_d)$. Let $\gamma = [E_d : E^\tau]$.

(B) There exists $\eta \in I_E$ such that $N_{E/E_d}(\eta) \equiv v_\tau^{|\tau|} \pmod{\mathfrak{q}}$. Let $\eta_i = \eta$ if $i = \gamma$ and let $\eta_i = 1$ otherwise.

(C) The natural I_E–module structure on $M(\mathfrak{p}^k) := \bigoplus_{\rho \in \mathcal{G}} \bigoplus_{1 \leq i \leq \gamma} I_E/\rho\tau^i\mathfrak{q}$ extends uniquely to an R–module structure such that for all

(15) $$A := \left(a_{\rho,i} + \rho\tau^i\mathfrak{q} : \rho \in \mathcal{G}, 1 \leq i \leq \gamma\right) \in M(\mathfrak{p}^k)$$

we have

(16) $$v_\tau A = \left(\tau(a_{\rho,i}\rho(\eta_i))v_\tau v_\rho v_\tau^{-1} v_\rho^{-1} + \rho\tau^{i+1}\mathfrak{q}\right)_{i,\rho}$$

and for all $\mu \in \mathcal{G}$,

(17) $$v_\mu A = \left(\mu(a_{\rho,i})v_\mu v_\rho v_{\mu\rho}^{-1} + \rho\mu\tau^i\mathfrak{q}\right)_{i,\rho}.$$

Moreover, with this R–module structure $M(\mathfrak{p}^k)$ is indecomposable with annihilator $\mathfrak{p}^k R$.

In Part (B) above N_{E/E_d} denotes the norm map from E to E_d. Similarly we have the norm map $N_{E/E^\tau} : E \longrightarrow E^\tau$. The choice of η in Part (B) corresponds to the choice of a and b in the construction of all Frobenius groups whose Frobenius complement is either the quaternion group (Example 1.3), a binary dihedral group (Lemma 14.3), or H_{120} (Lemma 14.7). As the proof of the above lemma will show, the group $\mathcal{G}$ and the group element $\langle \tau \rangle$ in Part (A) can be chosen almost canonically. $\mathcal{G}$ can also be chosen to have order at most 2. Indeed in all but three of the eleven cases of the proof of Theorem 12.6 $\mathcal{G}$ can be taken to be trivial, in which case the formula (16) simply says that $v_\tau A = \left(\tau(a_{\rho,i}\eta_i) + \tau^{i+1}\mathfrak{q}\right)_{i,\rho}$ and formula (17) says $v_\mu A = A$. The R–module $M(\mathfrak{p}^k)$ is determined up to isomorphism by $\mathfrak{p}^k$ independently of the above choices since it is indecomposable with annihilator $\mathfrak{p}^k R$.

Before giving the proof of the lemma we record as a corollary to the lemma and Theorem 13.2 the construction of all G–groups.

14.2. THEOREM. *The G-group corresponding to the orbit of any $\sum_{1\leq i\leq \rho}(\mathfrak{p}_i^{k_i})$ in $\mathcal{S}(G)$ under the bijection Ψ of Theorem 13.2 is the semidirect product*

$$(18) \qquad \left(\bigoplus_{1\leq i\leq \rho} M(\mathfrak{p}_i^{k_i})\right) \rtimes G.$$

We now prove Lemma 14.1.

PROOF. Let $f := [E : E_d]$ and let p be the unique rational prime in $\mathfrak{p}$. By Lemma 12.11 $\mathfrak{p}$ is unramified in $E/\mathcal{Z}$ and p does not divide f.

(A) When $\mathrm{Gal}(E/\mathcal{Z})$ is cyclic we can simply take $\mathcal{G} = 1$. When $\mathrm{Gal}(E/\mathcal{Z})$ is not cyclic, then it is generated by an element σ of odd order together with two distinct elements σ_0 and σ_1 of order 2 (cf. the proof of Theorem 12.6). Since $\mathrm{Gal}(E/E_d)$ is cyclic, it cannot contain $\langle \sigma_0, \sigma_1 \rangle$. If it is disjoint from $\langle \sigma_0, \sigma_1 \rangle$, then we can set $\tau = \sigma$ and let $\mathcal{G} = \langle \sigma_0, \sigma_1 \rangle$; finally if $\mathrm{Gal}(E/E_d)$ contains exactly one nontrivial element ρ of $\langle \sigma_0, \sigma_1 \rangle$, then we can let $\tau = \sigma\rho$ and let $\mathcal{G}$ be generated by any element of $\langle \sigma_0, \sigma_1 \rangle$ not in $\langle \rho \rangle$.

(B) By Theorem 12.6, $v_\tau^{|\tau|} \in \langle \zeta \rangle \subset I_E$. Since τ is just conjugation by v_τ, then $v_\tau^{|\tau|}$ is in E_d. Let $\mathfrak{a}$ denote the radical of $\mathfrak{q}$. Since norm maps for extensions of finite fields are surjective, then there exists $\eta' \in I_E$ with $N_{E/E_d}(\eta') \equiv v_\tau^{|\tau|} \pmod{\mathfrak{a}}$. Since $p \nmid f$, Hensel's Lemma implies that there exists $\eta'' \in E_d \cap I_E$ with $(\eta'')^f N_{E/E_d}(\eta') \equiv v_\tau^{|\tau|} \pmod{\mathfrak{q}}$. It suffices to set $\eta = \eta'\eta''$ since then $N_{E/E_d}(\eta) = (\eta'')^f N_{E/E_d}(\eta')$.

(C) Note that $|\tau| = f\gamma$. Since $\mathrm{Gal}(E/E_d)$ is contained in $\langle \tau \rangle$, it is generated by τ^γ, so $\tau^\gamma(\mathfrak{q}) = \mathfrak{q}$. Thus the value assigned to $v_\tau A$ in equation (16) is indeed in $M(\mathfrak{p}^k)$. Further, $\{\rho\tau^i : \rho \in \mathcal{G}, 1 \leq i \leq \gamma\}$ is a complete set of representatives for $\mathrm{Gal}(E/\mathcal{Z})/\mathrm{Gal}(E/E_d)$. Therefore $M(\mathfrak{p}^k)$ is isomorphic to $I_E/\mathfrak{p}^k I_E$, so its order is $|I_\mathcal{Z}/\mathfrak{p}^k|^{\deg \mathbb{Q}\langle G\rangle}$, which is the order of the indecomposable R-module with annihilator $\mathfrak{p}^k R$ (cf. Theorems 13.2 and 13.9). Let $\omega : I_E \longrightarrow \mathrm{End}_{I_\mathcal{Z}/\mathfrak{p}^k} M(\mathfrak{p}^k)$ be the natural multiplication map (recall that $M(\mathfrak{p}^k)$ is an I_E-module). Define V_τ and for each $\mu \in \mathcal{G}$ a map V_μ in $\mathrm{End}_{I_\mathcal{Z}/\mathfrak{p}^k} M(\mathfrak{p}^k)$ taking each A as in Equation (15) to the quantities in Equation (16) and Equation (17), respectively. Note that $\{v_\tau^i v_\mu : 0 \leq i < f\gamma, \mu \in \mathcal{G}\}$ is a basis for R as an I_E-module since $\{v_{\tau^i\mu} : 0 \leq i < f\gamma, \mu \in \mathcal{G}\}$ is a basis for R as an I_E-module and by Theorem 12.6 for each i and μ we have $v_\tau^i v_\mu = \delta v_{\tau^i\mu}$ for some unit $\delta \in I_E$. Hence we can define an additive map $\Omega : R \longrightarrow \mathrm{End}_{I_\mathcal{Z}/\mathfrak{p}^k} M(\mathfrak{p}^k)$ taking each $av_\tau^i v_\mu$ to $\omega(a) V_\tau^i V_\mu$ (for $a \in I_E, 0 \leq i < |\tau|$ and $\mu \in \mathcal{G}$). Suppose that $\mu, \delta \in \mathcal{G}$ and that $b \in I_E$; let A be as in Equation (15). Routine computations show that $\Omega(v_\mu b) = \Omega(v_\mu)\Omega(b)$ and that $\Omega(v_\tau b) = \Omega(v_\tau)\Omega(b)$. Similarly $\Omega(v_\mu)\Omega(v_\tau) = \Omega(v_\mu v_\tau)$ and $\Omega(v_\delta)\Omega(v_\mu) = \Omega(v_\delta v_\mu)$.

(Note that $\Omega(v_\delta v_\mu) = \Omega(v_\delta v_\mu v_{\delta\mu}^{-1})\Omega(v_{\delta\mu})$ and $\Omega(v_\mu v_\tau) = \Omega(v_\mu v_\tau v_\mu^{-1} v_\tau^{-1})\Omega(v_\tau v_\mu)$.)
We further claim that $\Omega(v_\tau^{|\tau|}) = \Omega(v_\tau)^{|\tau|}$. A straightforward computation shows that

$$V_\tau^\gamma(A) = \left(\tau^\gamma(a_{\rho,i})\tau^i\rho(\eta)\prod_{j=0}^{\gamma-1}\tau^j(v_\tau v_\rho v_\tau^{-1} v_\rho^{-1}) + \rho\tau^i\mathfrak{q} : \rho \in \mathcal{G}, 1 \leq i \leq \gamma\right)$$

(note that $\tau^j(\eta_{\gamma+i-j})$ is $\tau^i(\eta)$ if $i = j$ and is 1 otherwise). The product

$$N_{E/E^\tau}(v_\rho v_\tau v_\rho^{-1} v_\tau^{-1}) = \prod_{j=0}^{|\tau|-1} v_\tau^j v_\rho v_\tau v_\rho^{-1} v_\tau^{-1} v_\tau^{-j}$$

telescopes to $\rho(v_\tau^{|\tau|}) v_\tau^{-|\tau|}$. Thus by the choice of η we have modulo $\rho\mathfrak{q}$

$$\rho(N_{E/E_d}(\eta)) N_{E/E^\tau}(v_\tau v_\rho v_\tau^{-1} v_\rho^{-1}) \equiv \rho(v_\tau^{|\tau|})(\rho(v_\tau^{|\tau|}) v_\tau^{-|\tau|})^{-1} \equiv v_\tau^{|\tau|}.$$

One now can use the above formula for $V_\tau^\gamma(A)$ to show that as claimed $\Omega(v_\tau)^{|\tau|}(A) = V_\tau^{\gamma f}(A)$ is equal to

$$\left(a_{\rho,i}\tau^i(N_{E/E_d}(\rho\eta))N_{E/E^\tau}(v_\tau v_\rho v_\tau^{-1} v_\rho^{-1}) + \rho\tau^i\mathfrak{q}\right)_{\rho,i} = \Omega(v_\tau^{|\tau|})(A).$$

A tedious but straightforward computation using the formulas proved above shows that Ω preserves multiplication; this says that $M(\mathfrak{p}^k)$ has the required R-module structure.

It remains to show that $M(\mathfrak{p}^k)$ has annihilator $\mathfrak{p}^k R$ and is indecomposable as an R-module. Since $\mathfrak{p}^k \subset \mu\mathfrak{q}$ for all $\mu \in \text{Gal}(E/\mathcal{Z})$, then clearly the annihilator contains $\mathfrak{p}^k R$ and hence it has the form $\mathfrak{p}^j R$ for some $j \leq k$ (cf. Lemma 13.10(A)). Since p^j is in the annihilator, therefore

$$0 = p^j \left(1 + \rho\tau^i\mathfrak{q}\right)_{i,\rho} = \left(p^j + \rho\tau^i\mathfrak{q}\right)_{i,\rho},$$

so $p^j \in \mathfrak{q} \cap I_\mathcal{Z} = \mathfrak{p}^k$, so $j \geq k$. Thus $j = k$. Hence $\mathfrak{p}^k R$ is indeed the annihilator, and so $M(\mathfrak{p}^k)$ must be a sum of indecomposable R-modules whose annihilators contain $\mathfrak{p}^k R$, and at least one of these modules must have annihilator equal to $\mathfrak{p}^k R$. This one will then have the same order as $M(\mathfrak{p}^k)$, as was observed at the beginning of this proof of part (C). Hence $M(\mathfrak{p}^k)$ is in fact indecomposable. □

We now give a more elementary construction of the Frobenius groups whose Frobenius complement is the binary dihedral group

$$G = D_{4m} = \langle \zeta_{2m}, \mathbf{j} \rangle$$

where $m \geq 2$. We may identify R with the ring of real quaternions $\mathbb{Z}[1/2m, \zeta_{2m}, \mathbf{j}]$ (cf. Theorem 3.2); then $I_\mathcal{Z} = \mathbb{Z}[1/2m, \zeta_{2m} + \zeta_{2m}^{-1}]$. To see this one can argue directly or apply Theorem 5.2D when m is odd (whence G is a 1–complement with invariant $(4, m, \langle -1 \rangle)$) and apply Theorem 6.15 when m is even (whence G is a V_4–complement with reduced invariant $\langle (1, -1) \rangle$). It will be useful to keep in mind that

$$\left(\zeta_{2m} - \zeta_{2m}^{-1}\right)^2 = -4 + \left(\zeta_{2m} + \zeta_{2m}^{-1}\right)^2 \in I_\mathcal{Z}.$$

Suppose that $0 < k \in \mathbb{Z}$ and that $\mathfrak{p}$ is a maximal ideal of $I_\mathcal{Z}$. For each m, $\mathfrak{p}$ and k let us fix a choice of a and b as in the following lemma.

14.3. LEMMA. *There exists $a, b \in I_Z/\mathfrak{p}^k$ with $a^2 + b^2 = \left(\zeta_{2m} - \zeta_{2m}^{-1}\right)^2 + \mathfrak{p}^k$. There exists a unique $I_Z/\mathfrak{p}^k$-algebra isomorphism*

$$\theta : R/\mathfrak{p}^k R \longrightarrow M_2(I_Z/\mathfrak{p}^k)$$

mapping $(\zeta_{2m} - \zeta_{2m}^{-1}) + \mathfrak{p}^k R$ and $\mathbf{j} + \mathfrak{p}^k R$ to $A := \begin{bmatrix} a & b \\ b & -a \end{bmatrix}$ and $J := \begin{bmatrix} 0 & 1 \\ -1 & 0 \end{bmatrix}$, respectively.

Before proving this lemma, we state the main result for D_{4m}-groups.

Let $N(\mathfrak{p}^k)$ denote the set of 2×1 column matrices with entries from $I_Z/\mathfrak{p}^k$ considered as a left R-module (scalar multiplication by $r \in R$ is left multiplication by the matrix $\theta(r + \mathfrak{p}^k R)$); we may also consider the additive group $N(\mathfrak{p}^k)$ a D_{4m}-module via the injection $D_{4m} \longrightarrow R$.

14.4. THEOREM. *The Frobenius groups with Frobenius complement isomorphic to D_{4m} are exactly the groups isomorphic to a semidirect product of the form*

$$\left(\bigoplus_{i=1}^k N(\mathfrak{a}_i)\right) \rtimes D_{4m}$$

where $\sum_{i=1}^k \mathfrak{a}_i$ is in $\mathcal{S}(D_{4m})$.

The proofs of Lemma 14.3 and Theorem 14.4 will be combined.

PROOF. When $k = 1$, the existence of a and b in Lemma 14.3 follows from the fact that the quadratic form $x^2 + y^2$ is universal in any finite field of odd characteristic. The existence in the general case follows from the $k = 1$ case and the local squares theorem. (If $a_0^2 + b_0^2 \equiv \left(\zeta_{2m} - \zeta_{2m}^{-1}\right)^2 \pmod{\mathfrak{p}}$, then $\zeta_{2m} - \zeta_{2m}^{-1} - b_0^2$ is a square modulo $\mathfrak{p}$, and hence a square modulo any power of $\mathfrak{p}$. This argument assumes a_0 is a $\mathfrak{p}$-adic unit, which is true without loss of generality since $\zeta_{2m} - \zeta_{2m}^{-1}$ is a $\mathfrak{p}$-adic unit.) Since

$$\zeta_{2m} = \frac{1}{2}\left(\zeta_{2m} - \zeta_{2m}^{-1}\right) + \frac{1}{2}\left(\zeta_{2m} + \zeta_{2m}^{-1}\right),$$

the pair $1, \zeta_{2m} - \zeta_{2m}^{-1}$ is a basis for $\mathbb{Z}[1/2m, \zeta_{2m}]$ as an I_Z-module. Thus $1, \zeta_{2m} - \zeta_{2m}^{-1}$, $\mathbf{j}, \left(\zeta_{2m} - \zeta_{2m}^{-1}\right)\mathbf{j}$ is a basis for R as an I_Z-module. Hence

$$1 + \mathfrak{p}^k R, \ \zeta_{2m} - \zeta_{2m}^{-1} + \mathfrak{p}^k R, \ \mathbf{j} + \mathfrak{p}^k R, \ \left(\zeta_{2m} - \zeta_{2m}^{-1}\right)\mathbf{j} + \mathfrak{p}^k R$$

is a basis for $R/\mathfrak{p}^k R$ as an $I_Z/\mathfrak{p}^k$-module. Therefore there is a unique $I_Z/\mathfrak{p}^k$-linear function

$$\theta : R/\mathfrak{p}^k R \longrightarrow M_{\deg \mathbb{Q}\langle G \rangle}(I_Z/\mathfrak{p}^k)$$

taking the four elements of the above basis respectively to 1, A, J and AJ (here we have denoted by 1 the 2×2 identity matrix). Since

$$\det \begin{bmatrix} 1 & 0 & 1 & 0 \\ 0 & 1 & -1 & 0 \\ a & b & b & -a \\ -b & a & a & b \end{bmatrix} = 2(a^2 + b^2) \in (I_Z/\mathfrak{p}^k)^\bullet,$$

the matrices 1, A, J and AJ are linearly independent over $I_Z/\mathfrak{p}^k$ and hence the map θ is injective; it is surjective since its domain and codomain have the same number of elements (Lemma 13.10C). Finally, θ preserves multiplication since $A^2 = (a^2+b^2)\cdot 1$, $J^2 = -1$ and $AJ = -JA$. This completes the proof of the lemma. The theorem then follows from Theorem 13.2 and the construction in Proposition 13.8 of the Frobenius group associated with any $\sum_{i=1}^{k}(\mathfrak{a}_i) \in \mathcal{S}(D_{4m})$. □

14.5. REMARK. Note that $D_8 = \langle \mathbf{i}, \mathbf{j} \rangle$ is just the quaternion group. In this case $I_\mathbb{Z} = \mathbb{Z}[1/2]$ and the maximal ideals of $I_\mathbb{Z}$ are generated by the odd rational primes. For any odd prime power p^k we can identify the additive group $M(p^k)$ of Example 1.3 with $N\bigl((p\mathbb{Z}[\tfrac{1}{2}])^k\bigr)$. One checks that the actions of D_8 on $M(p^k)$ in Example 1.3 and Lemma 14.3 are identical. (Please note that in this case $\zeta_{2m} - \zeta_{2m}^{-1} = 2i$, so the a and b of Example 1.3 can be taken to be exactly half the a and b of Lemma 14.3.) Hence the groups constructed in Example 1.3 are well–defined and comprise all the Frobenius groups with Frobenius complement isomorphic to D_8 (note that all such groups have abelian kernel since D_8 has even order). The uniqueness assertion of Example 1.3 follows from the fact that $I_\mathbb{Z} = \mathbb{Z}[1/2]$ has no nontrivial automorphisms, so each orbit of $\mathcal{S}(D_8)/\operatorname{Aut} D_8$ is a singleton.

A corollary of Example 1.3 describes the orders of some Frobenius kernels.

14.6. COROLLARY. *The set of orders of all Frobenius kernels of Frobenius groups with Frobenius complement $\langle \mathbf{i}, \mathbf{j} \rangle$ is exactly the set of odd square integers larger than 1. The order of the Frobenius kernel of any Frobenius group whose Frobenius complement is not a $\mathbb{Z}$-group is an odd square.*

PROOF. The first sentence follows from the fact that the order of each indecomposable module $M(p^k)$ of Example 1.3 is the square p^{2k}. The second sentence follows from the first and the fact that by inspection (from Chapters 6 to 10) all Frobenius complements other than $\mathbb{Z}$–groups contain an isomorphic copy of $\langle \mathbf{i}, \mathbf{j} \rangle$, and hence the Frobenius kernels of Frobenius groups with such complements are also Frobenius kernels of Frobenius groups with complement $\langle \mathbf{i}, \mathbf{j} \rangle$. □

For the remainder of this chapter we let $G = H_{120}$ (cf. Chapter 3) and identify R and $I_\mathbb{Z}$ with $\mathbb{Z}[1/30, \sqrt{5}, \mathbf{i}, \mathbf{j}]$ and $\mathbb{Z}[1/30, \sqrt{5}]$, respectively (Theorem 3.2). Let $\mathbb{P}$ denote the set of all nontrivial powers $p^k \neq 1$ of rational primes $p > 5$. (As usual, p will be used only to denote a prime number.) The notation introduced in the next lemma will be used in our construction of all Frobenius groups with Frobenius complement $G = H_{120}$.

14.7. LEMMA. *Suppose $q = p^k \in \mathbb{P}$. Pick $a = a(q)$, $b = b(q)$ and $s = s(q)$ in $\mathbb{Z}_q$ with $a^2 + b^2 = -1$, and $30s = 1$. If $p \equiv 1$ or $4 \pmod{5}$, let $A = \mathbb{Z}_q$ and pick $c = c(q) \in A$ with $c^2 = 5 + q\mathbb{Z}$. If $p \equiv 2$ or $3 \pmod{5}$, let $A = \mathbb{Z}_q[X]/(Y)$ and $c = X + (Y)$ where $Y = X^2 - (5 + q\mathbb{Z})$. Then there are R-modules $M(q+)$ and $M(q-)$, both with additive group $A \oplus A$, such that for all $(\gamma, \delta) \in A \oplus A$,*

$$\frac{1}{30}(\gamma, \delta) = (s\gamma, s\delta), \quad \mathbf{i}(\gamma, \delta) = (a\gamma + b\delta, b\gamma - a\delta), \quad \mathbf{j}(\gamma, \delta) = (\delta, -\gamma),$$

and

$$\sqrt{5}(\gamma, \delta) = (c\gamma, c\delta) \text{ in } M(q+)$$

and

$$\sqrt{5}(\gamma, \delta) = (-c\gamma, -c\delta) \text{ in } M(q-).$$

Before proving the above lemma we state our main theorem on Frobenius groups with complement $G = H_{120}$. We need a bit more notation. Let $Free\mathbb{P}$ denote the set of all maps from $\mathbb{P}$ to the set of nonnegative integers with finite support (*i.e.*,

$Free\mathbb{P}$ is the free abelian monoid on the set $\mathbb{P}$). Let $\mathbb{P}_0$ denote the set of all proper prime powers p^k in $\mathbb{P}$ with $p \equiv 1$ or $4 \pmod 5$ and let $\mathbb{P}_1 = \mathbb{P} \setminus \mathbb{P}_0$. We give the set of all maps $\mathbb{P}_0 \longrightarrow \mathbb{Z}$ the lexicographic order (so $m > n$ if for some $q_0 \in \mathbb{P}_0$ we have $m(q_0) > n(q_0)$ and $m(q_1) = n(q_1)$ whenever $q_0 > q_1 \in \mathbb{P}_0$).

14.8. THEOREM. *For any m and n in $Free\mathbb{P}$, not both zero, the semidirect product*

(19) $$J(m,n) := \left(\oplus_{q \in \mathbb{P}}(M(q+)^{m(q)} \oplus M(q-)^{n(q)})\right) \rtimes H_{120}$$

is a Frobenius group with Frobenius complement isomorphic to H_{120}. Moreover any Frobenius group with Frobenius complement isomorphic to H_{120} is isomorphic to $J(m,n)$ for exactly one choice of m and n in $Free\mathbb{P}$, not both zero, with $m|\mathbb{P}_0 \geq n|\mathbb{P}_0$ and $n|\mathbb{P}_1 = 0$.

For example all Frobenius groups with complement H_{120} and order $11^4 \cdot 120$ are isomorphic to exactly one of the semidirect products

$$M(11+)^2 \rtimes G, \quad (M(11+) \oplus M(11-)) \rtimes G, \quad M(11^2+) \rtimes G.$$

Let us describe these groups more concretely. Since $11 \equiv 1 \pmod 5$ we have $M(11\pm) = \mathbb{Z}_{11}^2$ and $M(11^2+) = \mathbb{Z}_{121}^2$ and we can choose $a(11^i)$, $b(11^i)$, $c(11^i)$, $s(11^i)$ $(i = 1, 2)$ so that for all $(\gamma, \delta) \in \mathbb{Z}_{11}^2$

$$\frac{1}{30}(\gamma, \delta) = (7\gamma, 7\delta), \quad \sqrt{5}(\gamma, \delta) = (\pm 4\gamma, \pm 4\delta),$$
$$\mathbf{i}(\gamma, \delta) = (7\gamma + 4\delta, 4\gamma - 7\delta), \quad \mathbf{j}(\gamma, \delta) = (\delta, -\gamma)$$

and for all $(\gamma, \delta) \in \mathbb{Z}_{121}^2$,

$$\frac{1}{30}(\gamma, \delta) = (-4\gamma, -4\delta), \quad \sqrt{5}(\gamma, \delta) = (\pm 48\gamma, \pm 48\delta),$$
$$\mathbf{i}(\gamma, \delta) = (15\gamma + 4\delta, 4\gamma - 15\delta) \text{ and } \mathbf{j}(\gamma, \delta) = (\delta, -\gamma)$$

(the signs depend on whether we are in $M(11^i+)$ or $M(11^i-)$). The semidirect product $M(11+) \rtimes G$ is the H_{120}-group of minimal order; it is an easy exercise (at least with a table of primes less than 1000) to find the 92 isomorphism classes of H_{120}-groups with order at most $120 \cdot 10^6$. ($M(991+) \rtimes G$, which has order $120 \cdot 991^2$, is the one of maximum order.)

We have the following analogue to Corollary 14.6.

14.9. COROLLARY. *The set of natural numbers which are the orders of the Frobenius kernels of Frobenius groups with Frobenius complement H_{120} is the set S of all squares $n \neq 1$ relatively prime to 30 such that p_n is a fourth power for all primes $p \equiv 2$ or $3 \pmod 5$. The order of the Frobenius kernel of any nonsolvable Frobenius group is a member of S.*

PROOF. The first sentence follows from Theorem 14.8 and the fact that $|M(q\pm)|$ equals q^2 if $q \in \mathbb{P}_0$ and equals q^4 otherwise. The second sentence follows from the first and the fact that a Frobenius group is nonsolvable if and only if its Frobenius complement contains an isomorphic copy of H_{120}. □

We now prove Lemma 14.7 and Theorem 14.8.

PROOF. Theorem 13.2 describes the G–groups in terms of the maximal ideals of $I_{\mathcal{Z}}$ and their associated indecomposable R–modules. Each of these maximal ideals intersects down to a maximal ideal of $\mathbb{Z}[1/30]$, and hence to an ideal generated by a rational prime larger than 5. Now let us fix a proper prime power $q = p^k \in \mathbb{P}$. Let I_p denote the ring of p–adic integers. I_p contains a square root of 5 if and only if $\mathbb{Z}_q$ has an element with square $5 + q\mathbb{Z}$, and hence if and only if $p \equiv 1$ or 4 (mod 5) (by quadratic reciprocity $\left(\frac{5}{p}\right) = \left(\frac{p}{5}\right)$). In this case we let $\sqrt{5}$, as an element of I_p, denote the square root of 5 mapping to c (cf. Lemma 14.7) under the natural homomorphism $I_p \longrightarrow \mathbb{Z}_q$. Then in all cases we have a unique isomorphism

$$I_p[\sqrt{5}]/qI_p[\sqrt{5}] \longrightarrow \mathbb{Z}_q[c]$$

taking the coset of $\sqrt{5}$ to c. Let σ be the unique nontrivial automorphism of $I_{\mathcal{Z}}$ (so $\sigma(\sqrt{5}) = -\sqrt{5}$); σ is induced by an element of $\operatorname{Aut} G$ by Lemma 3.6. We have an embedding $\rho : I_{\mathcal{Z}} \longrightarrow I_p[\sqrt{5}]$ with $\rho(\sqrt{5}) = \sqrt{5}$. Since p is unramified in the extension $\mathbb{Q}[\sqrt{5}]/\mathbb{Q}$ [**R**, A2, p. 169], then

$$\mathfrak{p} = \mathfrak{p}(p) := \rho^{-1}(pI_p[\sqrt{5}]) \quad \text{and} \quad \sigma\mathfrak{p}$$

are exactly the extensions of $p\mathbb{Z}[1/30]$ to a maximal ideal of $I_{\mathcal{Z}}$ (they are equal if $p \equiv 2$ or 3 (mod 5)), and $\mathfrak{p}(q) := \mathfrak{p}(p)^k$ is equal to $\rho^{-1}(qI_p[\sqrt{5}])$ [**Re**, Theorem 5.1(ii), p. 68]. Thus there is an isomorphism

$$\theta_1 : I_{\mathcal{Z}}/\mathfrak{p}^k \longrightarrow \mathbb{Z}_q[c]$$

carrying the coset of $\sqrt{5}$ to c. We will lift θ_1 to an isomorphism on $R/(\mathfrak{p}R)^k$ essentially by tensoring with $\mathbb{Z}[\mathbf{i}, \mathbf{j}]$.

Since $\mathbb{Z}[\mathbf{i}, \mathbf{j}]$ is free as a $\mathbb{Z}$–module we have a natural isomorphism

(20) $$\mathbb{Z}[\mathbf{i}, \mathbf{j}] \otimes I_{\mathcal{Z}} \,/\, \mathbb{Z}[\mathbf{i}, \mathbf{j}] \otimes \mathfrak{p}^k \longrightarrow \mathbb{Z}[\mathbf{i}, \mathbf{j}] \otimes I_{\mathcal{Z}}/\mathfrak{p}^k \,.$$

The inclusion maps induce an isomorphism

$$\mathbb{Z}[\mathbf{i}, \mathbf{j}] \otimes I_{\mathcal{Z}} \longrightarrow R$$

which can be used to translate (20) into an isomorphism

$$R/R\mathfrak{p}^k \longrightarrow \mathbb{Z}[\mathbf{i}, \mathbf{j}] \otimes I_{\mathcal{Z}}/\mathfrak{p}^k \,.$$

Composition with $1 \otimes \theta_1$ gives an isomorphism

$$\theta_2 : R/(R\mathfrak{p})^k \longrightarrow \mathbb{Z}[\mathbf{i}, \mathbf{j}] \otimes \mathbb{Z}_q[c]$$

mapping the cosets of $1/30$, $\sqrt{5}$, $\mathbf{i}$ and $\mathbf{j}$ to $1 \otimes s$, $1 \otimes c$, $\mathbf{i} \otimes 1$ and $\mathbf{j} \otimes 1$, respectively.
There is an isomorphism of $\mathbb{Z}_q[c]$–algebras

$$\theta_3 : \mathbb{Z}[\mathbf{i}, \mathbf{j}] \otimes \mathbb{Z}_q[c] \longrightarrow M_2(\mathbb{Z}_q[c])$$

mapping $\mathbf{i} \otimes 1$ and $\mathbf{j} \otimes 1$ to

$$I := \begin{bmatrix} a & b \\ b & -a \end{bmatrix} \quad \text{and} \quad J := \begin{bmatrix} 0 & 1 \\ -1 & 0 \end{bmatrix},$$

respectively. After all, there is an isomorphism of $\mathbb{Z}_q[c]$–modules taking the basis 1, $\mathbf{i} \otimes 1$, $\mathbf{j} \otimes 1$, $\mathbf{ij} \otimes 1$ to the basis $1, I, J, IJ$ and this map preserves multiplication since $I^2 = J^2 = -1$ and $IJ = -JI$. The composition of θ_2 and θ_3 gives an isomorphism

$$R/(R\mathfrak{p})^k \longrightarrow M_2(\mathbb{Z}_q[c])\,.$$

14. CONCRETE CONSTRUCTIONS OF FROBENIUS GROUPS

It follows that $R\mathfrak{p}$ is a maximal ideal of R (consider the $k = 1$ case) and hence that $R/(R\mathfrak{p})^k$ is a finite primary principal ideal ring (Lemma 13.4). Thus as argued in the proof of Theorem 13.2

$$B := M_2(\mathbb{Z}_q[c]) \begin{bmatrix} 1 & 0 \\ 0 & 0 \end{bmatrix}$$

is an indecomposable R–module with annihilator $R\mathfrak{p}^k = (R\mathfrak{p})^k$. The isomorphism $B \longrightarrow A^2$ taking each $\begin{bmatrix} \gamma & 0 \\ \delta & 0 \end{bmatrix}$ in B to (γ, δ) puts an isomorphic R–module structure on A^2, which can be checked to be exactly the structure claimed for $M(q+)$. The automorphism σ of $I_{\mathcal{Z}}$ has an extension to an automorphism of R fixing $\mathbf{i}$ and $\mathbf{j}$; we will also denote this extension by σ. One checks that the scalar multiplication on $\sigma M(q+)$ is exactly that claimed for $M(q-)$; thus there is such an R–module and it is indecomposable with annihilator $\sigma(R\mathfrak{p}^k) = R\sigma(\mathfrak{p})^k$ (cf. Remark 13.6). This completes the proof of Lemma 14.7 and the calculation of the annihilators of the R–modules $M(q\pm)$ for $q \in \mathbb{P}$. From the proof of Proposition 13.8 it follows that all the groups $J(m, n)$ are H_{120}–groups. Note that given m and n in $Free\mathbb{P}$ (not both zero),

$$\psi(J(m, n)) = \sum_{q \in \mathbb{P}} \bigl(m(q)(\mathfrak{p}(q)) + n(q)(\sigma\mathfrak{p}(q))\bigr)$$

$$= \sum_{q \in \mathbb{P}_0} \bigl(m(q)(\mathfrak{p}(q)) + n(q)(\sigma\mathfrak{p}(q))\bigr) + \sum_{q \in \mathbb{P}_1} \bigl(m(q) + n(q)\bigr)(\mathfrak{p}(q))$$

$$= \sigma\left(\sum_{q \in \mathbb{P}_0} \bigl(n(q)(\mathfrak{p}(q)) + m(q)(\sigma\mathfrak{p}(q))\bigr) + \sum_{q \in \mathbb{P}_1} \bigl(m(q) + n(q)\bigr)(\mathfrak{p}(q))\right).$$

It follows that every H_{120}–group is isomorphic to one of the form $J(m, n)$ for some m and n (not both zero) in $Free\mathbb{P}$ (Proposition 13.8) and in fact for such m and n with $m|\mathbb{P}_0 \geq n|\mathbb{P}_0$ and $n|\mathbb{P}_1 = 0$ (Theorem 13.2). Now suppose $J(m, n)$ is isomorphic to $J(m', n')$ where $m|\mathbb{P}_0 \geq n|\mathbb{P}_0$, $m'|\mathbb{P}_0 \geq n'|\mathbb{P}_0$ and $n|\mathbb{P}_1 = n'|\mathbb{P}_1 = 0$. Then either $\psi(J(m, n))$ equals

$$\psi(J(m', n')) = \sum_{\sigma \in \mathbb{P}} m'(q)(\mathfrak{p}(q)) + n'(q)(\sigma\mathfrak{p}(q))$$

or it equals

$$\sigma\psi(J(m', n')) = \sum_{\sigma \in \mathbb{P}} n'(q)(\mathfrak{p}(q)) + m'(q)(\sigma\mathfrak{p}(q)).$$

In the first case $m = m'$ and $n = n'$. In the second $m|\mathbb{P}_0 = n'|\mathbb{P}_0$, $m'|\mathbb{P}_0 = n|\mathbb{P}_0$, $m|\mathbb{P}_1 = m'|\mathbb{P}_1$ and $n|\mathbb{P}_1 = n'|\mathbb{P}_1 = 0$. Then by hypothesis

$$m|\mathbb{P}_0 \geq n|\mathbb{P}_0 = m'|\mathbb{P}_0 \geq n'|\mathbb{P}_0 = m|\mathbb{P}_0,$$

so once again on all of $\mathbb{P}$ we have $m = m'$ and $n = n'$. This completes the proof of the uniqueness part of the theorem. $\square$

CHAPTER 15

Counting Frobenius Groups with Abelian Frobenius Kernel

Let G be a Frobenius complement and k be an integer, $k > 1$. Let $Iso(G,k)$ denote the set of isomorphism classes of Frobenius groups with Frobenius complement G and with abelian Frobenius kernel of order k. (The set $Iso(G)$ of Chapter 13 is the disjoint union of these sets $Iso(G,k)$.) In this chapter we show how the number of elements of $Iso(G,k)$ can be computed from the factorization $k = p_1^{a_1} \cdots p_\rho^{a_\rho}$ of k into a product of powers of distinct rational primes $p_1, \ldots, p_\rho$, together with the basic numerical invariants developed in Chapters 5 to 10 which determine the isomorphism class of G, namely, the core invariant Δ, the core index $[G:C]$, and if $[G:C] = 4, 12, 24$ or 120, the reduced invariant S. When $[G:C] = 1$ or 60, it is convenient to set S equal to the trivial subgroup of $\mathbb{Z}_m^\bullet \times \mathbb{Z}_n^\bullet$. We again use the notation of Chapter 12 and specifically that of Notation 12.1 and Remarks 12.5 and 12.9, so $E = \mathbb{Q}[\zeta] \supset \mathcal{Z}$.

Let $\mathcal{B}$ denote the image of $\text{Gal}(E/\mathcal{Z})$ under the natural isomorphism $\text{Aut}\, E \longrightarrow \mathbb{Z}_\mathfrak{f}^\bullet$. We now state a formula for $|Iso(G,k)|$ in terms of $\mathfrak{f}$, k, S, Δ and $\mathcal{B}$. In Remark 15.13 below we will show how to compute $\mathfrak{f}$ and $\mathcal{B}$ from Δ, $[G:C]$ and S. As in [**BrH**, p. 75] if $0 < v \in \mathbb{Z}$, we let $P(u,v)$ denote 0 if u is not a nonnegative integer and let it denote the coefficient of x^u in the power series expansion of $\prod_{i=1}^{\infty}(1-x^i)^{-v}$ otherwise. Also, if $i \leq \rho$ and $j \in \mathbb{Z}_\mathfrak{f}^\bullet$, then let $d(p_i, j)$ denote the subgroup of $\mathbb{Z}_\mathfrak{f}^\bullet$ generated by $\mathcal{B}$, j, and $p_i + \mathfrak{f}\mathbb{Z}$. Finally, let δ denote $(|S|, 3)$ and let $\mathcal{B}_0$ denote the kernel of the natural homomorphism $\mathbb{Z}_\mathfrak{f}^\bullet \longrightarrow \mathbb{Z}_{(n/t,\delta t)}^\bullet$.

15.1. THEOREM. *The number of elements of $Iso(G,k)$ is*

$$
\text{(21)} \qquad \frac{1}{|\mathcal{B}_0|} \sum_{j \in \mathcal{B}_0} \prod_{i=1}^{\rho} P\left(\frac{a_i}{d(p_i,j)}, \frac{\phi(\mathfrak{f})}{d(p_i,j)}\right).
$$

The reader can consult [**BrH**, Lemma 11.12 and Remark 11.13C, pp. 77–78] for a combinatorial interpretation of $P(u,v)$ as a kind of partition function and for effective methods of computing it. The next remark indicates that the appearance of a partition function in this context is to be expected.

15.2. REMARK. Consider the case that G is the cyclic group $\langle \zeta_n \rangle$, $n > 1$. Then $\mathfrak{f} = n$ and $E = \mathbb{Q}[\zeta_n] = \mathcal{Z}$, so $\mathcal{B}$ is trivial and $\mathcal{B}_0 = \mathbb{Z}_n^\bullet$. Then the formula (21) becomes

$$
\frac{1}{\phi(n)} \sum_{j \in \mathbb{Z}_n^\bullet} \prod_{i=1}^{\rho} P\left(\frac{a_i}{d(p_i,j)}, \frac{\phi(n)}{d(p_i,j)}\right)
$$

which is the formula in [**BrH**, Theorem 11.7, p. 75] for the number of isomorphism classes of metabelian Frobenius groups with kernel and complement of orders k and n, respectively. Actually the formula of [**BrH**] also applies when $n = 1$ to give that the number of isomorphism classes of abelian groups of order k is

$$\prod_{i=1}^{\rho} P(a_i, 1);$$

this standard result [**H**, Corollary, p. 115] follows from the fundamental theorem of finite abelian groups.

15.3. EXAMPLE. The following table, constructed using the above theorem, Remark 15.13 and Example 13.11, gives the number of isomorphism classes of non-metabelian Frobenius groups J with abelian Frobenius kernel subject to restrictions on the core index I of a Frobenius complement of J and on the order $|J|$ of J.

	$\|J\| \leq 10^5$	$\|J\| \leq 5 \times 10^5$	$\|J\| \leq 10^6$
$I = 1$	100	268	395
$I = 4$	151	422	605
$I = 12$	23	59	84
$I = 24$	6	18	25
$I = 60$	2	8	12
$I = 120$	0	0	1
TOTAL	282	775	1122

The number of metabelian Frobenius groups is discussed in Remark 11.13(A) of [**BrH**] (there are 568,220 with order at most one million).

The above calculations were very much simplified by some easy consequences of the fact that we were only considering Frobenius groups of order at most one million; this implies, for example, that if the Frobenius group has core index greater than one, then in fact the core is cyclic, and if the core index is one, then the degree of the rational truncated group ring of the Frobenius complement is at most 4.

The proof of Theorem 15.1 will proceed in a sequence of lemmas. In these lemmas $\mathcal{Z}^{\operatorname{Aut} G}$ will denote the subfield of $\mathcal{Z}$ fixed under all the automorphisms of $\mathcal{Z}$ induced by automorphisms of G. (Each automorphism of G induces an automorphism of $\mathbb{Q}\langle G \rangle$ and hence an automorphism of its center $\mathcal{Z}$. Note that $\mathcal{Z}$ itself is the fixed ring of $\mathbb{Q}\langle G \rangle$ under the inner automorphisms of G.) Our first task is to calculate $\mathcal{Z}^{\operatorname{Aut} G}$. We begin with a slight refinement of part of Lemma 6.8.

15.4. LEMMA. *Suppose that* $\sigma \in \operatorname{Aut} G$ *and* $\sigma(x) = y^c x^h$ *where* $c \in \mathbb{Z}_m$ *and* $h \in \mathbb{Z}_n^\bullet$. *Then* $h \equiv 1 \pmod{\delta t}$.

PROOF. Suppose $\delta \neq 1$, since otherwise the lemma follows trivially from Lemma 6.8. Then $[G : C] = 12$, 3 divides n (so $3 \cdot 3_t$ divides n), and S is nontrivial. We have $\sigma(y) = y^b$ for some $b \in \mathbb{Z}_m^\bullet$. Suppose u, v, x, y, z is an r–sequence for G with invariant (a, g). Then $u_1 := \sigma(u)$, $v_1 := \sigma(v)$, $x_1 := \sigma(x)$, $y_1 := \sigma(y)$, $z_1 := \sigma(z)$ is also an r–sequence for G with the same invariant. Proceeding as in the proof of Lemma 7.15 we set $z_1 = \gamma x^f z^j$ where $\gamma \in \langle u, v, y \rangle$, j is 1 or 2, and $f \in \mathbb{Z}_n$. Since

$$y^{ab} = \sigma(zyz^{-1}) = x^f z^j y^b z^{-j} x^{-f} = y^{ba^j r^f},$$

therefore $1 = a^{j-1}r^f$. Next, modulo $\langle u, v, y \rangle$ we have

$$x^{h(n//3)} \equiv \sigma(x^{n//3}) \equiv \sigma(z^3) \equiv (x^f z^j)^3 \equiv x^{f(1+g^j+g^{2j})} x^{j(n//3)}.$$

Hence

(22) $$h(n//3) \equiv f(1 + g^j + g^{2j}) + j(n//3) \pmod{n}.$$

One more such computation. Again modulo $\langle y, u, v \rangle$ we have

$$x^{gh} \equiv \sigma(zxz^{-1}) \equiv x^f z^j x^h z^{-j} x^{-f} \equiv x^{hg^j}$$

so $g^{j-1} \equiv 1 \pmod{n}$.

Now suppose $j = 2$. Then by the last congruence, g is trivial. Since $S \neq 1$, therefore 3 divides $|a|$. Since $1 = a^{j-1}r^f$, then the order of a divides the order of r, namely t. Hence $3|t$. Since $a^3 = r^{n//3}$ (Lemma 7.8), then the order of a is $3 \cdot 3_t$ (note 3_t is the order of $r^{n//3}$), which does not divide t. This contradiction shows $j = 1$. Hence $r^f = 1$, so $f \equiv 0 \pmod{3_t}$. By Lemma 7.8 $g \equiv 1 \pmod{3}$, so $1 + g + g^2 \equiv 0 \pmod{3}$. Hence by the congruence (22)

$$(n//3)(h-1) \equiv f(1+g+g^2) \equiv 0 \pmod{3 \cdot 3_t},$$

so $h \equiv 1 \pmod{3 \cdot 3_t}$ as claimed. □

15.5. LEMMA. *There exists an r–sequence x, y for the core of G such that for every $d \in \mathbb{Z}_m^{\bullet}$ and $e \in \mathbb{Z}_n^{\bullet}$ with $e \equiv 1 \pmod{\delta t}$, there exists an automorphism $\tau \in \operatorname{Aut} G$ with $\tau(y) = y^d$ and $\tau(x) = x^e$.*

PROOF. The argument breaks into several cases. Write $e = q + n\mathbb{Z}$ where $q \in \mathbb{Z}$, so $q \equiv 1 \pmod{([G:C], 2)t}$.

Case 1: $[G:C] = 1$. Since $e \equiv 1 \pmod{t}$, then for any r–sequence x, y, we have $x^e y^d x^{-e} = (y^d)^r$. It follows that there is an endomorphism of G taking x, y to x^e, y^d. Since $G = \langle x^e, y^d \rangle$, this endomorphism is the required automorphism.

Case 2: $[G:C] = 4$. Let u, v, x, y be an r–sequence for G with invariant (a, g, c, h). There is an endomorphism of G mapping u, v, x, y to u, v^q, x^e, y^d respectively; this map is surjective and hence is the required automorphism. (The point here is that u, v^q, x^e, y^d is an r–sequence for G with the same invariant as u, v, x, y and hence is a set of generators for G satisfying the same relations as u, v, x, y. For example one computes that since q is odd, $h^q = h$ and hence $v^q x^e v^{-q} = x^{eh^q} = (x^e)^h$.)

Case 3: $[G:C] = 12$ and $3 \nmid n$. We use the notation of Theorem 7.11. By the first case there is an automorphism τ of $\langle x_0, y_0 \rangle$ taking x_0, y_0 to x_0^e, y_0^d. Then the automorphism $1 \otimes \tau$ of $G := H_{24} \times J$ maps $x = (-1, x_0)$ and $y = (1, y_0)$ to x^e and y^d, as required.

In the next three cases we have $[G:C] = 12$ and we let u, v, x, y, z be an r–sequence for G with invariant (b, h).

Case 4: $[G:C] = 12$, 3 divides n, $S = 1$ and $e \equiv 1 \pmod{3}$. Since $S = 1$, therefore $b = 1$ and $h = 1$ (cf. Lemma 7.8), so z commutes with x and y. One checks that u, v, x^e, y^d, z^q is an r–sequence for G with invariant (b, h). (That $z^q uz^{-q} = vu$ and $z^q vz^{-q} = u$ follows from the hypothesis that $e \equiv 1 \pmod{3}$.) Then by Lemma 7.9 G has an automorphism taking x and y to x^e and y^d.

Case 5: $[G:C] = 12$, 3 divides n, $S = 1$ and $e \not\equiv 1 \pmod{3}$. As in the previous case (b, h) is trivial. Since 3 divides n, therefore $q \equiv 2 \pmod{3}$. Hence $-vu, -v$,

x^e, y^d, z^q is an r-sequence for G with trivial invariant. Hence by Lemma 7.9 G admits an automorphism taking x, y to x^e, y^d.

Case 6: $[G:C] = 12$, 3 divides n, and $S \neq 1$. Then $\delta = 3$, so $e \equiv 1 \pmod{3t}$. As in the previous case, it suffices to show u, v, x^e, y^d, z^q is an r-sequence with invariant (b,h). There exists $i \in \mathbb{Z}$ with $q = 1 + 3it$, so by Lemma 7.8

$$b^q = b(b^3)^{ti} = b(r^{n//3})^{ti} = b;$$

hence $z^q y^d z^{-q} = (y^d)^{b^q} = (y^d)^b$. The rest is routine.

In the next three cases we have $[G:C] = 24$ and we let u, v, w, x, y, z be an r-sequence for G with invariant (b,h).

Case 7: $[G:C] = 24$ and $3 \nmid n$. By Proposition 8.6 it suffices to show that u, v, w, x^e, y^d, z is also an r-sequence for G with invariant (b,h), and this can be verified by routine computations. (Recall that $z^3 = x^{n//3} = 1$ since 3 does not divide n and that z commutes with x and y.)

Case 8: $[G:C] = 24$, $3|n$, and $e \equiv 1 \pmod 3$. In this case it suffices to verify that u, v, w, x^e, y^d, z^q is an r-sequence for G with invariant (b,h). In Case 4 we saw that u, v, x^e, y^d, z^q was an r-sequence for $\langle u,v,x,y,z\rangle$. The rest is transparent except for the equality $wz^q w^{-1} = (x^e)^{-n//3} v(z^q)^2$. To verify this write $q = 1 + 3i$ (where $i \in \mathbb{Z}$) and recall that $h \equiv -1 \pmod{3_n}$ by Lemma 8.4. Note that

$$-(n//3) + ih(n//3) \equiv -(n//3) - 3i(n//3) + 2i(n//3) \pmod n.$$

Hence

$$wz^q w^{-1} = wzx^{i(n//3)}w^{-1} = x^{-n//3}vz^2 x^{ih(n//3)}$$
$$= x^{-n//3}x^{-3i(n//3)}vz^2 x^{2i(n//3)} = (x^e)^{-n//3}v(z^q)^2.$$

Case 9: $[G:C] = 24$, $3|n$ and $e \equiv 2 \pmod 3$. In this case it suffices to observe that $-v$, $-u$, wz^2, x^e, y^d, z^q is an r-sequence for G with invariant (b,h). Since $h \equiv -1 \pmod{3_n}$, then $h(n//3) \equiv -n//3 \pmod n$. Writing $q = 2 + 3i$ we therefore have

$$(wz^2)(z^q)(wz^2)^{-1} = wz^2 x^{i(n//3)} w^{-1}$$
$$= (x^{-n//3}vz^2)^2 x^{ihn//3} = (x^{2+3i})^{-(n//3)}(-u)x^{(n//3)2i}z^4$$
$$= (x^e)^{-n//3}(-u)(z^q)^2.$$

The remaining verifications are routine.

Case 10: $[G:C] = 60$. We use the notation of Theorem 9.1 and assume $G = J \times H_{120}$. As in the first case J has an automorphism τ taking x_0 and y_0 to x_0^q and y_0^d, respectively. Then the automorphism $\tau \times 1$ of $J \times H_{120}$ maps $x = (x_0, -1)$ and $y = (y_0, 1)$ to x^e and y^d, respectively.

Case 11: $[G:C] = 120$. Let μ be the automorphism of $\mathbb{Z}\langle J \times H_{120}\rangle$ induced by the automorphism $\tau \times 1$ of the previous case. Then without loss of generality G is the S_5-complement constructed in Theorem 10.6 and μ commutes with ρ. This implies that μ induces an automorphism of G fixing $\hat\rho$. □

15.6. LEMMA. $\mathcal{Z}^{\operatorname{Aut} G} \subset L$.

PROOF. We use the notation and the division into cases of the proof of Theorem 12.6. In Cases 1, 3, 4, 6 and 8 we have $L \supset \mathcal{Z}$, so there is nothing to prove. We now consider the remaining cases. For any $\tau \in \operatorname{Aut} G$ we also let τ denote the automorphism of $\mathbb{Q}\langle G\rangle$ that it induces.

Case 2: $[G:C] = 4$ and $|S| \le 2$. There exists $\tau \in \operatorname{Aut} G$ mapping an r–sequence u, v, x, y to $u, -v, x, y$. Thus $\mathcal{Z}^{\operatorname{Aut} G} \subset \mathcal{Z}^\tau \subset L[v]^\tau = L$. Here $\mathcal{Z}^\tau$ has denoted the set of elements of $\mathcal{Z}$ fixed by the automorphism of $\mathcal{Z}$ induced by τ.

Case 5: $[G:C] = 12$, $3|n$ and $S = 1$. If u, v, x, y, z is an r–sequence for G, then u, v, x, y, z^{1+n} is an r–sequence for G with the same trivial invariant. Hence there exists $\tau \in \operatorname{Aut} G$ mapping the first r–sequence to the second. Hence $\tau(Az) = Az^{n+1} \ne Az$ since $z^n = (x^{n//3})^{n/3} \ne 1$. Since Az satisfies a cubic polynomial over L we have
$$\mathcal{Z}^{\operatorname{Aut} G} \subset \mathcal{Z}^\tau \subset L[Az]^\tau = L.$$

Case 7: $[G:C] = 24$, $3 \nmid n$, $S = 1$. If u, v, w, x, y, z is an r–sequence for G, then $u, v, -w, x, y, z$ is an r–sequence with the same invariant, so some $\tau \in \operatorname{Aut} G$ maps the first r–sequence onto the second. Then $\tau(\eta) = -\eta$, so $\mathcal{Z}^{\operatorname{Aut} G} \subset \mathcal{Z}^\tau \subset L[\eta]^\tau = L$.

Case 9: $[G:C] = 24$ and $3|n$. Let u, v, w, x, y, z be an r–sequence for G. Note that
$$\begin{aligned} wz^{n+1}w^{-1} = (vz^{-1})^{n+1} &= vz^{-1}(vz^{-1}vz^{-1}vz^{-1})^{n/3} \\ &= vz^{-1}(z^{-3})^{n/3} = v(z^{n+1})^{-1}. \end{aligned}$$
It follows easily that u, v, w, x, y, z^{n+1} is an r–sequence for G with the same invariant; hence there exists $\tau \in \operatorname{Aut} G$ mapping the first to the second. As in Case 5 we then have $\mathcal{Z}^{\operatorname{Aut} G} \subset L[Az]^\tau = L$.

Case 10: $[G:C] = 60$. If ψ_0 is the automorphism of Lemma 3.6A, then $1 \times \psi_0$ is an automorphism of $G = J \times H_{120}$; the induced automorphism of $\mathbb{Q}\langle G \rangle$ maps $\sqrt{5}$ to $-\sqrt{5}$. Thus
$$\mathcal{Z}^{\operatorname{Aut} G} \subset \left(\mathbb{Q}[\zeta_{mn/t}] \otimes \mathbb{Q}[\sqrt{5}] \right)^{1 \times \psi_0} \subset L.$$

Case 11: $[G:C] = 120$. The automorphism $1 \times \psi_0$ of $J \times H_{120}$ in the previous case extends to the automorphism $1 \otimes \rho_1$ of $\mathcal{E} = \mathbb{Q}\langle J \times H_{120}\rangle$, which commutes with ρ and hence induces an automorphism τ of G fixing $\widehat{\rho}$. Thus
$$\mathcal{Z}^{\operatorname{Aut} G} \subset \left(\mathbb{Q}[\zeta_{mn/t}] \otimes \mathbb{Q}[\sqrt{5}] \right)^\tau \subset L.$$
$\square$

15.7. THEOREM. $\mathcal{Z}^{\operatorname{Aut} G}$ *is a cyclotomic extension of* $\mathbb{Q}$ *generated by a primitive* $(n/t, \delta t)^{\text{th}}$ *root of unity.*

PROOF. We let x, y be as in Lemma 15.5, and we identify x^t with $\zeta_{n/t}$ and y with ζ_m, so that $L = \mathbb{Q}[\zeta_{mn/t}]$ (cf. Remark 12.5). We begin by justifying the inclusions
$$(23) \qquad \mathbb{Q}[\zeta_{(n/t, \delta t)}] \subset \mathcal{Z}^{\operatorname{Aut} G} \subset L^{\operatorname{Aut} G} \subset L^H$$
where H is the image of the canonical homomorphism
$$\theta : \mathbb{Z}_m^\bullet \times \ker\left(\mathbb{Z}_n^\bullet \longrightarrow \mathbb{Z}_{\delta t}^\bullet\right) \longrightarrow \operatorname{Aut} L$$
which for each $a \in \mathbb{Z}_m^\bullet$ and $g \in \ker\left(\mathbb{Z}_n^\bullet \longrightarrow \mathbb{Z}_{\delta t}^\bullet\right)$ has
$$\theta(a,g)(\zeta_m) = \zeta_m^a \quad \text{and} \quad \theta(a,g)(\zeta_{n/t}) = \zeta_{n/t}^g.$$

First suppose $\tau \in \operatorname{Aut} G$. Then by Lemmas 6.8 and 15.4, $\tau(y) = y^a$ and $\tau(x) = y^b x^h$ where $a \in \mathbb{Z}_m^\bullet$, $b \in \mathbb{Z}_m$ and $h \in \ker\left(\mathbb{Z}_n^\bullet \longrightarrow \mathbb{Z}_{\delta t}^\bullet\right)$. In all cases δt

divides n (as noted earlier, $\delta \neq 1$ only if 3 divides n, cf. Remark 7.4). Since $h \equiv 1 \pmod{\delta t}$,

$$hn/(n/t, \delta t) = h(n/\delta t)(\delta t/(n/t, \delta t))$$
$$\equiv (n/\delta t)(\delta t/(n/t, \delta t)) \equiv n/(n/t, \delta t) \pmod{n}.$$

Hence by Lemma 6.7

$$\begin{aligned}\tau(\zeta_{(n/t,\delta t)}) &= \tau(x^{n/(n/t,\delta t)}) \\ = (y^b x^h)^{n/(n/t,\delta t)} &= x^{hn/(n/t,\delta t)} \\ = x^{n/(n/t,\delta t)} &= \zeta_{(n/t,\delta t)}.\end{aligned}$$

Thus justifies the first inclusion of the display (23). The second inclusion follows from Lemma 15.6; and the third is immediate from Lemma 15.5. It now suffices to show that

$$\phi(mn/t) = |H|\phi((n/t, \delta t)),$$

since this will imply that the inclusions of display (23) are all equalities.

We next show that the sequence

$$1 \longrightarrow 1 \times \ker\left(\mathbb{Z}_n^\bullet \to \mathbb{Z}_{[n/t,\delta t]}^\bullet\right) \xrightarrow{\psi} \mathbb{Z}_m^\bullet \times \ker\left(\mathbb{Z}_n^\bullet \to \mathbb{Z}_{\delta t}^\bullet\right) \xrightarrow{\theta} H \longrightarrow 1$$

is exact, where ψ is the inclusion map. (Recall that n is a multiple of δt and hence is a multiple of the least common multiple $[n/t, \delta t]$ of n/t and δt.) If $g \in \ker\left(\mathbb{Z}_n^\bullet \to \mathbb{Z}_{[n/t,\delta t]}^\bullet\right)$, then $\theta((1, g))(y) = y$ and $\theta((1, g))(x^t) = x^{tg} = x^t$ since $g \equiv 1 \pmod{n/t}$ by hypothesis. This shows that the image of ψ is contained in the kernel of θ. Now suppose $(a, g) \in \ker \theta$. Then $y^a = y$ and $x^{tg} = x^t$, so $a = 1$ and $g \equiv 1 \pmod{n/t}$. By the choice of g we have $g \equiv 1 \pmod{\delta t}$, so $g \equiv 1 \pmod{[n/t, \delta t]}$. Thus (a, g) is in the image of ψ. This proves the exactness of the above sequence. We conclude that

$$|H|\phi(n)/\phi([n/t, \delta t]) = \phi(m)\phi(n)/\phi(\delta t).$$

Hence it suffices to show that

$$\phi(n/t)\phi(\delta t) = \phi([n/t, \delta t])\phi((n/t, \delta t)).$$

Recall that for any positive integer z we let z_0 denote the product of the distinct rational prime divisors of z, so that

(24) $$\phi(z) = z\phi(z_0)/z_0$$

[**NZ**, Theorem 2.16, p. 48]. Let $q = (n//t)_0$. Since t_0 divides n/t (Lemma 5.3(A)), we have $[n/t, \delta t]_0 = t_0 q = (n/t)_0$ (recall that if $\delta \neq 1$ then $\delta = 3$ divides n). Also note that $(n/t, \delta t)_0 = (\delta t)_0$. Therefore

$$\begin{aligned}&\phi([n/t, \delta t])\phi((n/t, \delta t)) \\ &= \frac{[n/t, \delta t]\phi([n/t, \delta t]_0)}{[n/t, \delta t]_0} \frac{(n/t, \delta t)\phi((n/t, \delta t)_0)}{(n/t, \delta t)_0} \\ &= \frac{(n/t)\phi((n/t)_0)}{(n/t)_0} \frac{\delta t\,\phi((\delta t)_0)}{(\delta t)_0} \\ &= \phi(n/t)\phi(\delta t).\end{aligned}$$

This completes the proof of Theorem 15.7. $\square$

We now turn more directly toward the proof of Theorem 15.1. The argument will parallel and generalize that of [**BrH**, §11]. Let $\mathcal{Z}_1$ be a field lying between E and $\mathcal{Z}$, and set $\mathcal{G} = \mathrm{Gal}(\mathcal{Z}_1/\mathcal{Z}^{\mathrm{Aut}\,G})$. For each $\sigma \in \mathcal{G}$ let $I_\sigma = \mathcal{Z}^\sigma \cap I_\mathcal{Z}$. We also let P_σ denote the set of nonzero primary ideals of I_σ and let $\mathcal{S}_\sigma$ denote the free abelian semigroup on the set P_σ. For example, if σ fixes $\mathcal{Z}$, then $I_\sigma = I_\mathcal{Z}$, $P_\sigma = P$ and $\mathcal{S}_\sigma = \mathcal{S}(G)$ (cf. the first paragraph of Chapter 13). In general I_σ, P_σ and $\mathcal{S}_\sigma$ depend only on the restriction of σ to $\mathcal{Z}$. If $\mathfrak{q} \in P_\sigma$, let

$$P(\mathfrak{q}) = \{\mathfrak{p} \in P : \mathfrak{p} \cap I_\sigma = \mathfrak{q}\}.$$

Finally, if $\eta \in \mathcal{S}_\sigma$, let

$$\eta \otimes I_\mathcal{Z} = \sum_{\mathfrak{p} \in P} \eta(\mathfrak{p} \cap I_\sigma)\mathfrak{p} \in \mathcal{S}(G).$$

(We are viewing elements of $\mathcal{S}_\sigma$ as functions from P_σ to the set of nonnegative integers having nontrivial finite support.)

15.8. LEMMA. *For all $\sigma \in \mathcal{G}$ and $\mathfrak{p} \in P$*

$$P(\mathfrak{p} \cap I_\sigma) = \{\tau\mathfrak{p} : \tau \in \langle\sigma\rangle\}.$$

PROOF. Let $\mathfrak{a}$ and $\mathfrak{b}$ be maximal ideals of $I_\mathcal{Z}$ and let i and j be positive integers. By Lemma 12.11 no maximal ideal of I_σ ramifies in $I_\mathcal{Z}$. Thus $\mathfrak{a}^i \cap I_\sigma = (\mathfrak{a} \cap I_\sigma)^i$. Therefore, if $\mathfrak{a}^i \cap I_\sigma = \mathfrak{b}^j \cap I_\sigma$, then $i = j$ and $\mathfrak{a} \cap I_\sigma = \mathfrak{b} \cap I_\sigma$, so $\tau(\mathfrak{a}) = \mathfrak{b}$ for some $\tau \in \mathrm{Gal}(\mathcal{Z}/\mathcal{Z}^\sigma) = \langle\sigma|\mathcal{Z}\rangle$. Hence, if $\mathfrak{b}^j \in P(\mathfrak{a}^i \cap I_\sigma)$, then $\mathfrak{b}^j \in \{\tau(\mathfrak{a}^i) : \tau \in \langle\sigma\rangle\}$. Thus $P(\mathfrak{a}^i \cap I_\sigma) \subset \{\tau(\mathfrak{a}^i) : \tau \in \langle\sigma\rangle\}$; the reverse inclusion is transparent. $\square$

15.9. THEOREM. *There exists a $|\mathcal{G}| : 1$ covering of $\mathcal{S}(G)/\mathrm{Aut}\,G$ by the set $\mathcal{U} := \bigcup_{\sigma \in \mathcal{G}} \{\sigma\} \times \mathcal{S}_\sigma$ mapping each pair (σ, η) to the orbit of $\eta \otimes I_\mathcal{Z}$.*

PROOF. By Galois theory both $\mathcal{G} = \mathrm{Gal}(\mathcal{Z}_1/\mathcal{Z}^{\mathrm{Aut}\,G})$ and $\mathrm{Aut}\,G$ induce the same set of automorphisms of $\mathcal{Z}$. Hence the natural actions of $\mathcal{G}$ and $\mathrm{Aut}\,G$ on $\mathcal{S}(G)$ have exactly the same set of orbits. Now pick $\sigma \in \mathcal{G}$ and $\gamma \in \mathcal{S}(G)$. Let H denote the stabilizer of γ with respect to the action of $\mathcal{G}$ on $\mathcal{S}(G)$; then $|[\gamma]| = [\mathcal{G} : H]$. Since $\mathcal{G}$ is abelian, H is also the stabilizer of any element of $[\gamma]$. We now show $\sigma \in H$ if and only if $\gamma = \eta \otimes I_\mathcal{Z}$ for some $\eta \in \mathcal{S}_\sigma$. First, if $\gamma = \eta \otimes I_\mathcal{Z}$ for some $\eta \in \mathcal{S}_\sigma$, then

$$\begin{aligned}\sigma(\gamma) &= \sum_{\mathfrak{p} \in P} \eta(\sigma^{-1}(\mathfrak{p}) \cap I_\sigma)\sigma(\sigma^{-1}(\mathfrak{p})) \\ &= \sum_{\mathfrak{p} \in P} \eta(\mathfrak{p} \cap I_\sigma)\mathfrak{p} = \gamma,\end{aligned}$$

so $\sigma \in H$. Next suppose $\sigma \in H$. Then

$$\sum_{\mathfrak{p} \in P} \gamma(\sigma^{-1}(\mathfrak{p}))\mathfrak{p} = \sum_{\mathfrak{p} \in P} \gamma(\mathfrak{p})\mathfrak{p},$$

so

$$\gamma(\mathfrak{p}) = \gamma(\sigma^{-1}(\mathfrak{p})) = \gamma(\sigma^{-2}(\mathfrak{p})) = \ldots.$$

Thus for each $\mathfrak{p} \in P$, the function γ is constant on

$$\{\tau(\mathfrak{p}) : \tau \in \langle\sigma\rangle\} = P(\mathfrak{p} \cap I_\sigma)$$

(Lemma 15.8). Hence we can unambiguously define $\eta_\sigma \in \mathcal{S}_\sigma$ by setting $\eta_\sigma(\mathfrak{p} \cap I_\sigma) = \gamma(\mathfrak{p})$ for all $\mathfrak{p} \in P$. Clearly, η_σ is the unique element of $\mathcal{S}_\sigma$ with $\gamma = \eta_\sigma \otimes I_\mathcal{Z}$, as claimed. The $|H|$ elements (σ, η_σ) of $\mathcal{U}$ (where σ ranges over H) are exactly the elements (τ, η) of $\mathcal{U}$ such that $\gamma = \eta \otimes I_\mathcal{Z}$. Since $[\gamma]$ has $[\mathcal{G} : H]$ elements, there are exactly $[\mathcal{G} : H]|H| = |\mathcal{G}|$ elements of $\mathcal{U}$ mapping to $[\gamma]$. □

15.10. LEMMA. *Suppose $(\sigma, \eta) \in \mathcal{U}$. The order of the Frobenius kernel of the Frobenius group associated with $[\eta \otimes I_\mathcal{Z}] \in \mathcal{S}(G)/\operatorname{Aut} G$ by the map Ψ of Theorem 13.2 is*

$$\Gamma_\eta := \prod_{\mathfrak{q} \in P_\sigma} |I_\sigma/\mathfrak{q}|^{\eta(\mathfrak{q})[\mathcal{Z}:\mathcal{Z}^\sigma]\deg \mathbb{Q}\langle G\rangle}.$$

PROOF. By Theorem 13.9 the Frobenius kernel, call it M, of the Frobenius group associated with the orbit of

$$\eta \otimes I_\mathcal{Z} = \sum_{\mathfrak{q} \in P_\sigma} \sum_{\mathfrak{p} \in P(\mathfrak{q})} \eta(\mathfrak{q})\mathfrak{p}$$

has order

$$\prod_{\mathfrak{q} \in P_\sigma} \prod_{\mathfrak{p} \in P(\mathfrak{q})} |I_\mathcal{Z}/\mathfrak{p}|^{\eta(\mathfrak{q})\deg \mathbb{Q}\langle G\rangle}.$$

Consider any $\mathfrak{q} \in P_\sigma$ and $\mathfrak{p} \in P(\mathfrak{q})$. Write $\mathfrak{p} = \mathfrak{a}^i$ where $\mathfrak{a}$ is a maximal ideal of $I_\mathcal{Z}$. Since $\mathfrak{a}$ is unramified in $\mathcal{Z}/\mathcal{Z}^\sigma$ we have

$$\mathfrak{q} = \mathfrak{p} \cap I_\sigma = \mathfrak{a}^i \cap I_\sigma = (\mathfrak{a} \cap I_\sigma)^i$$

and $[\mathcal{Z} : \mathcal{Z}^\sigma] = fg$ where f is the residue class degree of $\mathfrak{a}$ in $\mathcal{Z}/\mathcal{Z}^\sigma$ and

$$\begin{aligned} g &= |\{\mathfrak{b} \in P : \mathfrak{b} \text{ is maximal and } \mathfrak{b} \cap I_\sigma = \mathfrak{a} \cap I_\sigma\}| \\ &= |\{\mathfrak{b}^i \in P : \mathfrak{b} \text{ is maximal and } \mathfrak{b}^i \cap I_\sigma = \mathfrak{q}\}| \\ &= |P(\mathfrak{q})|. \end{aligned}$$

Note that f depends only on $\mathfrak{a} \cap I_\sigma$ and hence on $\mathfrak{q}$. Therefore

$$|I_\mathcal{Z}/\mathfrak{p}| = |I_\mathcal{Z}/\mathfrak{a}|^i = |I_\sigma/I_\sigma \cap \mathfrak{a}|^{if} = |I_\sigma/\mathfrak{q}|^f$$

so that

$$\begin{aligned} |M| &= \prod_{\mathfrak{q} \in P_\sigma} |I_\sigma/\mathfrak{q}|^{|P(\mathfrak{q})|f\eta(\mathfrak{q})\deg \mathbb{Q}\langle G\rangle} \\ &= \prod_{\mathfrak{q} \in P_\sigma} |I_\sigma/\mathfrak{q}|^{[\mathcal{Z}:\mathcal{Z}^\sigma]\eta(\mathfrak{q})\deg \mathbb{Q}\langle G\rangle} = \Gamma_\eta. \end{aligned}$$

□

We continue to use the notation of the previous lemma.

15.11. LEMMA. *Let $\sigma \in \mathcal{G}$. Then*

$$|\{\eta \in \mathcal{S}_\sigma : \Gamma_\eta = k\}| = \prod_{i=1}^\rho P\left(\frac{a_i}{f(p_i)[\mathcal{Z}:\mathcal{Z}^\sigma]\deg \mathbb{Q}\langle G\rangle}, g(p_i)\right)$$

where for each rational prime p we let $f(p)$ denote the residue class degree of p in $\mathcal{Z}^\sigma/\mathbb{Q}$ and $g(p)$ denote the number of maximal ideals of I_σ containing p.

PROOF. For each rational prime p let $\Delta(p)$ denote the set of maximal ideals of I_σ containing p, so that $g(p) = |\Delta(p)|$. Let $\eta \in \mathcal{S}_\sigma$. Then

$$\eta = \sum_p \sum_{\mathfrak{a} \in \Delta(p)} \sum_{j=1}^\infty \eta(\mathfrak{a}^j) \mathfrak{a}^j$$

where the first sum is over all rational primes p not dividing $|G|$. Then

$$\Gamma_\eta = \prod_p \prod_{\mathfrak{a} \in \Delta(p)} \prod_{j=1}^\infty |I_\sigma/\mathfrak{a}^j|^{\eta(\mathfrak{a}^j)[\mathcal{Z}:\mathcal{Z}^\sigma]\deg \mathbb{Q}\langle G \rangle}$$

by the previous lemma. Since $|I_\sigma/\mathfrak{a}^j| = p^{jf(p)}$, therefore

$$\Gamma_\eta = \prod_p \prod_{\mathfrak{a}} \prod_j p^{jf(p)\eta(\mathfrak{a}^j)[\mathcal{Z}:\mathcal{Z}^\sigma]\deg \mathbb{Q}\langle G \rangle} = \prod_p p^{h(p)}$$

where

$$h(p) = \sum_{\mathfrak{a} \in \Delta(p)} \sum_{j=1}^\infty jf(p)\eta(\mathfrak{a}^j)[\mathcal{Z}:\mathcal{Z}^\sigma]\deg \mathbb{Q}\langle G \rangle .$$

Hence $\Gamma_\eta = k = \prod_{i=1}^\rho p_i^{a_i}$ if and only if $\eta(\mathfrak{a}^j) = 0$ whenever a maximal ideal $\mathfrak{a}$ of I_σ contains a rational prime equal to none of the p_i ($i \leq \rho$) and for all $i \leq \rho$

$$a_i = f(p_i)[\mathcal{Z}:\mathcal{Z}^\sigma]\deg \mathbb{Q}\langle G \rangle \sum_{\mathfrak{a} \in \Delta(p_i)} \sum_{j=1}^\infty j\eta(\mathfrak{a}^j) .$$

But for each $i \leq \rho$ the number of solutions in integers $x_{j,\mathfrak{a}}$ (where $\mathfrak{a} \in \Delta(p_i)$ and $j \geq 1$) of the equation

$$\frac{a_i}{f(p_i)[\mathcal{Z}:\mathcal{Z}^\sigma]\deg \mathbb{Q}\langle G \rangle} = \sum_{\mathfrak{a} \in \Delta(p_i)} \sum_{j=1}^\infty jx_{j,\mathfrak{a}}$$

is exactly

$$P\left(\frac{a_i}{f(p_i)[\mathcal{Z}:\mathcal{Z}^\sigma]\deg \mathbb{Q}\langle G \rangle}, g(p_i)\right)$$

(cf. [**BrH**, Lemma 11.12C, p. 77]). Hence the number of $\eta = \sum_{i=1}^\rho \sum_{\mathfrak{a} \in \Delta(p_i)} \sum_{j=1}^\infty \eta(\mathfrak{a}^j)\mathfrak{a}^j$ with $\Gamma_\eta = k$ is exactly

$$\prod_{i=1}^\rho P\left(\frac{a_i}{f(p_i)[\mathcal{Z}:\mathcal{Z}^\sigma]\deg \mathbb{Q}\langle G \rangle}, g(p_i)\right) .$$

$\square$

15.12. THEOREM. *With $f(p_i)$ and $g(p_i)$ as in Lemma 15.11,*

$$|Iso(G,k)| = \frac{1}{|\mathcal{G}|} \sum_{\sigma \in \mathcal{G}} \prod_{i=1}^\rho P\left(\frac{a_i}{f(p_i)[\mathcal{Z}:\mathcal{Z}^\sigma]\deg \mathbb{Q}\langle G \rangle}, g(p_i)\right) .$$

PROOF. Let Γ_η be as in the previous two lemmas. By Theorems 13.2 and 15.9 there exists a $|\mathcal{G}| : 1$ cover of $Iso(G)$ by $\mathcal{U}$, so

$$|\mathcal{G}||Iso(G,k)| = |\{(\sigma,\eta) \in \mathcal{U} : \Gamma_\eta = k\}|\,.$$

The previous lemma then implies that

$$|Iso(G,k)| = \frac{1}{|\mathcal{G}|} \sum_{\sigma \in \mathcal{G}} \prod_{i=1}^{\rho} P\left(\frac{a_i}{f(p_i)[\mathcal{Z}:\mathcal{Z}^\sigma] \deg \mathbb{Q}\langle G\rangle}, g(p_i)\right)\,.$$

$\square$

The above theorem takes a very natural form when we pick $\mathcal{Z}_1 = \mathcal{Z}$, so $\mathcal{G} = \text{Gal}(\mathcal{Z}/\mathcal{Z}^{\text{Aut }G})$ and $[\mathcal{Z}:\mathcal{Z}^\sigma] = |\sigma|$. In the following proof of Theorem 15.1, however, we will take $\mathcal{Z}_1 = E$.

PROOF. Let $\mathcal{Z}_1 = E$, so $\mathcal{G} = \text{Gal}(E/\mathcal{Z}^{\text{Aut }G})$. By Theorem 15.7 the image of $\mathcal{G}$ under the natural isomorphism $\text{Aut } E \longrightarrow \mathbb{Z}_\mathfrak{f}^\bullet$ is exactly

$$\mathcal{B}_0 = \ker \mathbb{Z}_\mathfrak{f}^\bullet \longrightarrow \mathbb{Z}_{(n/t,\delta t)}^\bullet$$

where $\delta = (3, |S|)$. We now prove that if $i \leq \rho$ and if $\sigma \in \mathcal{G}$ corresponds to $j \in \mathcal{B}_0$, then

$$f(p_i)[\mathcal{Z}:\mathcal{Z}^\sigma]\deg \mathbb{Q}\langle G\rangle = d(p_i, j)$$

and hence

$$d(p_i,j)g(p_i) = [\mathcal{Z}:\mathcal{Z}^\sigma]f(p_i)g(p_i)\deg \mathbb{Q}\langle G\rangle$$
$$= [E:\mathcal{Z}][\mathcal{Z}:\mathcal{Z}^\sigma][\mathcal{Z}^\sigma:\mathbb{Q}] = \phi(\mathfrak{f})\,.$$

Theorem 15.1 will then follow immediately from Theorem 15.12. The order of $\mathcal{B}$ is $\deg \mathbb{Q}\langle G\rangle$, and $[\mathcal{Z}:\mathcal{Z}^\sigma]$ is the order of the coset $\sigma \text{Gal}(E/\mathcal{Z})$ in

$$\text{Gal}(E/\mathcal{Z}^{\text{Aut }G})/\text{Gal}(E/\mathcal{Z}) \cong \text{Gal}(\mathcal{Z}/\mathcal{Z}^{\text{Aut }G})\,.$$

Hence $[\mathcal{Z}:\mathcal{Z}^\sigma]$ is the order of $j\mathcal{B}$ in $\mathcal{B}_0/\mathcal{B}$. Finally $f(p_i)$ is the order of $(p_i+\mathfrak{f}\mathbb{Z})\langle\mathcal{B},j\rangle$ in

$$\mathbb{Z}_\mathfrak{f}^\bullet/\langle\mathcal{B},j\rangle \cong (\mathbb{Z}_\mathfrak{f}^\bullet/\mathcal{B})/(\langle\mathcal{B},j\rangle/\mathcal{B})$$
$$\cong \text{Gal}(\mathcal{Z}/\mathbb{Q})/\text{Gal}(\mathcal{Z}/\mathcal{Z}^\sigma) \cong \text{Gal}(\mathcal{Z}^\sigma/\mathbb{Q})\,.$$

Thus $[\mathcal{Z}:\mathcal{Z}^\sigma]f(p_i)\deg\mathbb{Q}\langle G\rangle$ is the order of the subgroup of $\mathbb{Z}_\mathfrak{f}^\bullet$ generated by $\mathcal{B}$, j, and $p+\mathfrak{f}\mathbb{Z}$, i.e., $d(p_i,j)$. This completes the proof of Theorem 15.1. $\square$

15.13. REMARK. The reduction to elementary number theory of the computation of $|Iso(G,k)|$ requires a description of how to compute $\mathfrak{f}$ and $\mathcal{B}$ in terms of the numerical invariants Δ, $[G:C]$ and S for G. Both $\mathfrak{f}$ and $\text{Gal}(E/\mathcal{Z})$ are computed in the proof of Theorem 12.6. In particular if we write $\mathfrak{f} = smn/t$, then we have $s = 1, 2, 1, 2, 6, 2, 4, 2, 6, 5, 5$ in the Cases 1 through 11, respectively. By inspection then

$$\begin{aligned} s &= (5, [G:C]) \quad \text{if either } [G:C] = |S| = 4 \text{ or } [G:C] \in \{1, 60, 120\}; \\ s &= 4 \quad \text{if } [G:C] = 24, |S| = 1 \text{ and } 3 \nmid n; \text{ and} \\ s &= 2(3//|S|, n, [G:C]) \quad \text{otherwise.} \end{aligned}$$

The description of $\mathcal{B}$ in terms of Δ, $[G:C]$ and S is a bit more complicated. Let

$$(a,g,c,h) \in \mathbb{Z}_m^\bullet \times \mathbb{Z}_n^\bullet \times \mathbb{Z}_m^\bullet \times \mathbb{Z}_n^\bullet$$

denote $(1,1,1,1)$ if $[G:C] = 1$ or 60; let it be an invariant for G if $[G:C] = 4$; and let it be such that $c = 1$, $h = 1$ and (a,g) is an invariant for G otherwise. Such a 4-tuple is easily constructed from $[G:C]$, Δ and S (cf. Theorems 6.14, 7.13A, 8.7A and 10.6A) and except when $[G:C]$ is 4 or 12 is unique. Let $q = 5$ if $[G:C] \geq 60$ and otherwise let q denote $[G:C]_0$, the product of the distinct prime divisors of $[G:C]$. Let $\mu = 2^{120/[G:C]}$ if $[G:C] \geq 60$, $\mu = (-1)^{4/[G:C]}$ if $[G:C] \leq 4$, and $\mu = (3_\mathfrak{f})^2 + 8^{3_\mathfrak{f} \cdot 24/[G:C]}$ if $[G:C] = 12$ or 24. By the Chinese Remainder Theorem there exist unique r^*, g^* and h^* in $\mathbb{Z}_\mathfrak{f}^\bullet$ with $r^* \equiv r$ (mod m); $r^* \equiv 1$ (mod $\mathfrak{f}/m$); $g^* \equiv a$ (mod m); $g^* \equiv g$ (mod $(\mathfrak{f}/m)//q$); $g^* \equiv \mu$ (mod $q_\mathfrak{f}$); and, finally, $h^* \equiv c$ (mod m) and $h^* \equiv h$ (mod $\mathfrak{f}/m$) if $[G:C] = 4$, and $h^* \equiv (-1)^{[G:C]+1}$ (mod $2_\mathfrak{f}$) and $h^* \equiv 1$ (mod $\mathfrak{f}//2$) if $[G:C] \neq 4$. We now claim that in all cases,

(25) $$\mathcal{B} = \langle r^*, g^*, h^* \rangle.$$

To prove formula (25), one simply translates the case-by-case descriptions of a set of generators for $\mathrm{Gal}(E/\mathcal{Z})$ in the proof of Theorem 12.6 into a set of generators for $\mathcal{B}$ using the following simple principle. *Suppose that $\tau \in \mathrm{Aut}\,\mathbb{Q}[\zeta_\mathfrak{f}]$ and that $\mathfrak{f}$ is the product of three pairwise relatively prime positive integers b_1, b_2, b_3. For each $i \leq 3$, let η_i be a b_i^{th} root of unity in $\mathbb{Q}[\zeta_\mathfrak{f}]$ and suppose $\tau(\eta_i) = \eta_i^{e_i}$. If $e \in \mathbb{Z}$ is a solution to the set of congruences $x \equiv e_i$ (mod b_i) (for $i = 1,2,3$), then $e + \mathfrak{f}\mathbb{Z}$ is the image of τ under the natural isomorphism $\mathrm{Aut}\,\mathbb{Q}[\zeta_\mathfrak{f}] \longrightarrow \mathbb{Z}_\mathfrak{f}^\bullet$.* We illustrate the use of this fact to calculate $\mathcal{B}$ in two of the cases from the proof of Theorem 12.6 and leave the remaining cases to the reader.

Case 6: $[G:C] = 12$, $3|n$ and $|S| > 1$. In this case $s = 2$, $q = 6$, $g \equiv 1$ (mod $(6t)_n$) (by Lemma 7.8), $6_\mathfrak{f} = 3_{n/t} \cdot 4$, (a,g) is an invariant for G with respect to an r-sequence u, v, x, y, z, and by Euler's theorem

$$\mu = 3_\mathfrak{f}^2 + 8^{3\phi(3_\mathfrak{f})} \equiv 1 \pmod{6_\mathfrak{f}}.$$

We factor $\mathfrak{f}$ in three ways

$$m \cdot (\mathfrak{f}/m) = (2_\mathfrak{f}) \cdot (\mathfrak{f}//2) = m \cdot 6_{\mathfrak{f}/m} \cdot (n/t)//6.$$

The above factors are the orders, respectively, of the elements

$$y, \quad vx^t, \quad v, \quad yx^{2t}, \quad y, \quad x^{t(n//3)}v, \quad x^{t(6_n)}.$$

$\mathrm{Gal}(E/\mathcal{Z})$ is generated by σ, σ_1 and τ_1. Since $\sigma(y) = y^r$ and $\sigma(vx^t) = vx^t$, then σ maps to r^*. Since $\sigma_1(v) = v^{-1}$ and $\sigma_1(yx^{2t}) = yx^{2t}$, then σ_1 maps to h^*. Note $g \equiv 1$ (mod 3_n), so $gt(n//3) \equiv t(n//3)$ (mod n), and hence τ_1 fixes $x^{t(n//3)}v$. Also $\tau_1(y) = y^a$ and $\tau_1(x^{t6_n}) = (x^{t6_n})^g$, so τ_1 maps to g^*. Thus $\mathrm{Gal}(E/\mathcal{Z})$ maps to $\langle r^*, g^*, h^* \rangle$.

Case 11: $[G:C] = 120$. Then $\mu = 2$, $s = q = 5$ and $\mathfrak{f} = 5mn/t$. We factor $\mathfrak{f}$ in two ways:

$$m \cdot \mathfrak{f}/m = m \cdot n/t \cdot 5.$$

The factors of $\mathfrak{f}$ above are respectively the orders of

$$\zeta_m \otimes 1, \quad \zeta_{n/2t} \otimes -\beta, \quad \zeta_m \otimes 1, \quad \zeta_{n/2t} \otimes -1, \quad 1 \otimes \beta.$$

Gal$(E/\mathcal{Z})$ is generated by σ and ρ_3. Since $\sigma(\zeta_m\otimes 1) = (\zeta_m\otimes 1)^r$ and $\sigma(\zeta_{n/2t}\otimes-\boldsymbol{\beta}) = \zeta_{n/2t}\otimes-\boldsymbol{\beta}$, σ is mapped to r^*. On the other hand $\rho_3(\zeta_m\otimes 1) = \rho(\zeta_m\otimes 1) = (\zeta_m\otimes 1)^a$, $\rho_3(\zeta_{n/2t}\otimes-1) = \rho(\zeta_{n/2t}\otimes-1) = (\zeta_{n/2t}\otimes-1)^g$ and by equation (11) in Chapter 12
$$\rho_3(1\otimes\boldsymbol{\beta}) = (1\otimes i\boldsymbol{\alpha}^{-1})(1\otimes\rho_1(\boldsymbol{\beta}))(1\otimes i\boldsymbol{\alpha}^{-1})^{-1} = (1\otimes\boldsymbol{\beta})^2.$$
Thus ρ_3 maps to g^*. Since $2_{\mathfrak{f}} = 2$, then $h^* = 1$. Hence $\mathcal{B}$ is indeed generated by r^*, g^* and h^*.

CHAPTER 16

Isomorphism Invariants for Frobenius Complements

We now study combinations of isomorphism invariants of Frobenius complements which determine those groups up to isomorphism. Our object is to prove Theorem 1.6 of the Introduction, and to investigate the extent to which a Frobenius complement G is determined by its truncated group ring $\mathbb{Z}_{(G)}\langle G\rangle$. For A either a group or ring, let $[A]$ be the isomorphism class of A (in the appropriate sense).

We continue to use the notation of Notation 12.1 and Remark 12.5, so that in particular G will denote a Frobenius complement with core C and core invariant (m, n, T). We let S denote the signature of G. Set $k = |C'\mathcal{Z}(C)|$; thus $k = mn/t$ and $L \cong \mathbb{Q}[\zeta_k]$ (cf. Remark 12.5).

16.1. THEOREM. $[G]$ *is determined by any of the following four collections of data:*

(D1) $|C|$ and $[\mathbb{Z}_{(G)}\langle G\rangle]$;
(D2) $[G : C]$ and $[\mathbb{Z}_{(G)}\langle G\rangle]$;
(D3) $|G|$, $|C|$, and $[\mathcal{Z}]$;
(D4) $|G|$, $|C|$, and S.

Theorem 1.6 is of course equivalent to the assertion that $[G]$ is determined by the data in (D4). We begin the proof of Theorem 16.1 with four lemmas. The first simply assembles some material that can be read off of Theorems 5.2D, 6.15, 7.11, 7.13C, 8.7C, 9.1, and 10.6C.

16.2. LEMMA. *For each of the six possible values of* $[G : C]$, *the corresponding values of* $\deg \mathbb{Q}\langle G\rangle/t$ *and* $[\mathbb{Q}\langle G\rangle : \mathbb{Q}]/\phi(mn)$ *are as indicated below.*

$[G : C] =$	1	4	12	24	60	120
$[\mathbb{Q}\langle G\rangle : \mathbb{Q}]/\phi(mn) =$	1	4	(n, 3)4	(n, 3)8	8	16
$\deg \mathbb{Q}\langle G\rangle/t =$	1	2 or 4	2 or 2(n,3)	2 or 4	2	4

Our second lemma relates the integer units in $\mathbb{Z}_{(G)}\langle G\rangle$ to the order of G. Its proof involves a (somewhat implicit) construction of bases for integral truncated group rings of Frobenius complements as $\mathbb{Z}$–modules.

16.3. LEMMA. *Let p be a rational prime. Then $p \in \mathbb{Z}_{(G)}\langle G\rangle^\bullet$ if and only if p divides the order of G.*

PROOF. First suppose that as a $\mathbb{Z}$–module $\mathbb{Z}\langle G\rangle$ has a basis of the form $\gamma_1, \gamma_2, \ldots, \gamma_s$ where $\gamma_1 = 1$. This is also a basis for $\mathbb{Z}_{(G)}\langle G\rangle$ considered as a $\mathbb{Z}_{(G)}$–module. If p is a unit in $\mathbb{Z}_{(G)}\langle G\rangle$, then for some integers a_i and b_i with each b_i

dividing some power of $|G|$ we have
$$p((a_1/b_1)\gamma_1 + \cdots + (a_s/b_s)\gamma_s) = 1\gamma_1$$
so $pa_1/b_1 = 1$. Thus p divides $|G|$. On the other hand if p divides $|G|$ then p is a unit in $\mathbb{Z}_{(G)}\langle G \rangle$ by the definition of the truncated group ring.

We complete the proof of the lemma by verifying the supposition of the first sentence of the proof. If G is cyclic of order n, then $\mathbb{Z}\langle G \rangle \cong \mathbb{Z}[\zeta_n]$, which has basis $1, \zeta, \ldots, \zeta^{\phi(n)-1}$. Lemma 3.5 implies that $B := \{1, \mathbf{i}, \mathbf{j}, \boldsymbol{\alpha}\}$ is a basis for $\mathbb{Z}[H_{24}] \cong \mathbb{Z}\langle H_{24} \rangle$ and that $B + \beta B$ is a basis for $\mathbb{Z}[H_{120}] \cong \mathbb{Z}\langle H_{120} \rangle$. Thus the supposition is valid in these three cases; it follows that it is valid in *all* cases since in Chapters 5 through 10 we constructed all integral truncated group rings of Frobenius complements from those of these three cases by the processes of forming tensor products and of forming the "cyclic extensions" of Albert's Theorem 2.5; both these constructions clearly preserve the property of having a basis including the element 1. □

The next lemma shows how the signature S of G determines the reduced invariant (if there is one) and core invariant of G. Recall that if the core index of G is 4, 12, 24, or 120, then invariants (and reduced invariants) for G, usually associated with r–sequences of G, have been defined (cf. Definitions 6.4, 7.2, 8.1 and 10.1).

16.4. LEMMA. *Let* $\theta : \mathbb{Z}_m^{\bullet} \times \mathbb{Z}_n^{\bullet} \longrightarrow \mathbb{Z}_{mn}^{\bullet}$ *be the canonical isomorphism and set* $T_1 = \theta(T \times \langle 1 + n\mathbb{Z} \rangle)$. *If* $[G : C] = 4, 12, 24$ *or* 120 *select an invariant for* G. *Define*

$$\begin{aligned} B &= \langle 1 + mn\mathbb{Z} \rangle & \text{if } [G:C] = 1 \text{ or } 60; \\ &= \langle \theta(a,g), \theta(b,h) \rangle & \text{if } [G:C] = 4 \text{ and} \\ & \quad (a,g,b,h) \text{ is the selected invariant for G;} \\ &= \langle \theta(a,g) \rangle & \text{if } [G:C] = 12, 24, \text{ or } 120 \end{aligned}$$

and (a, g) *is the selected invariant for* G.

Then S *is generated by* T_1 *and* B. *Moreover* T_1 *is exactly the set of elements of* S *of order dividing* t. *If* $G = C$ *then* B *is trivial; if* 3 *divides* $|S|$ *and* $[G : C] = 12$ *then* S *is cyclic of order* $3t$ *and* B *is the Sylow 3–subgroup of* S; *and in all other cases* B *is the Sylow 2–subgroup of* S.

PROOF. T_1 is the image of C under the map $G \longrightarrow \mathbb{Z}_{mn}^{\bullet}$ of Definition 1.4, so $T_1 \subset S$. If $[G : C] = 1$ or 60 then $T_1 = S$ since G is generated by C together with some elements of G commuting with every element of C (cf. Theorem 9.1 for the case $[G : C] = 60$). Now suppose $[G : C] = 4$ and u, v, x, y is an r-sequence for G with invariant (a, g, b, h). Then S is generated by T_1 and the images of u and v in S, namely $\theta(a, g)$ and $\theta(b, h)$, respectively. Thus S is generated by B and T_1; the set of elements of S of order dividing the odd number t is T_1; and B is the Sylow 2-subgroup of S. Next suppose $[G : C] = 12$. Say u, v, x, y, z is an r-sequence for G having the selected invariant (a, g). Since u and v commute with x and y, S is generated by T_1 and the image of $\langle z \rangle$ in S, namely $B = \langle (a, g) \rangle$. First consider the case that 3 does not divide $|S|$. Then (a, g) is trivial, so $3 \nmid t$, $B = 1$, and $S = T_1$. Next suppose 3 divides $|S|$ but not t. Then $a^3 = r^{n//3} = 1$, so (a, g) has order exactly 3. That is, T_1 has order prime to 3 and B has order 3. Finally suppose 3 divides t. Since $a^3 = r^{n//3} \neq 1$, then (a, g) has order $3(3t)$ and $\theta^{-1}(S) = \langle (r, 1), (a, g) \rangle = \langle (r^{3t}, 1), (a, g) \rangle$. Since $|r^{3t}| = t//3$ is relatively prime to

$|(a,g)| = 3(3_t)$, then S is cyclic of order $3t$. One easily checks now in all cases with $[G : C] = 12$ that T_1 is the set of elements of S of order dividing t and B is the Sylow 3–subgroup of S.

Next consider the case that $[G : C] = 24$. If u, v, w, x, y, z is an r–sequence with the selected invariant (a, g), then we have $B = \langle \theta(a, g) \rangle$, which is the image of w in S. Since u, v and z commute with x and y, this says S is generated by T_1 and B. Since t is odd and $|B|$ a 2–power, then B is the Sylow 2–subgroup of S and T_1 the set of elements of S of order dividing t. Finally suppose that $[G : C] = 120$. We may assume our group G and selected invariant (a, g) are those constructed in Theorem 10.6. Then S is again generated by T_1 and the image of $\widehat{\rho}$ in S, namely $B = \langle (a, g) \rangle$. And again T_1 is the set of elements of S of order dividing t and B is the Sylow 2–subgroup. □

The next lemma relates the signature S to the center $\mathcal{Z}$ of $\mathbb{Q}\langle G \rangle$.

16.5. LEMMA. *The diagram*

$$\begin{array}{ccc} G & \longrightarrow & \mathbb{Z}_{mn}^{\bullet} \\ \delta \downarrow & & \downarrow \rho \\ \mathrm{Aut}\, L & \longrightarrow & \mathbb{Z}_k^{\bullet} \end{array}$$

commutes, where the top map is that of Definition 1.5, the bottom map is the canonical isomorphism, the right-hand map is the canonical one, and the left-hand map is induced by conjugation by elements of G (if $g \in G$ and $\eta \in L$, then $\delta(g)(\eta) = \overline{g}\eta\overline{g}^{-1}$). Moreover the image of δ is $\mathrm{Gal}(L/L \cap \mathcal{Z})$ and the restriction of ρ to S is injective.

Before proving this lemma we state a corollary which begins to suggest how special are S (when compared to other subgroups of $\mathbb{Z}_{mn}^{\bullet}$) and $\mathcal{Z} \cap L$ (when compared to other subfields of cyclotomic extensions of $\mathbb{Q}$).

16.6. COROLLARY. *S is isomorphic to $\mathrm{Gal}(L/L \cap \mathcal{Z})$, and both have a cyclic subgroup of index at most 2. (S is not cyclic if and only if G is a V_4-complement with reduced invariant of order 4.)*

The isomorphism of the corollary is immediate from Lemma 16.5; the rest of the corollary follows from the analysis of S in Lemma 16.4.

We now prove Lemma 16.5.

PROOF. Let x, y be an r–sequence for C, so

$$\langle x^t y \rangle = \langle x^t, y \rangle = \mathcal{Z}(C)C'$$

is a normal subgroup of G and $L = \mathbb{Q}[\overline{x^t y}]$. Then conjugation by elements of G induces automorphisms of $\langle x^t y \rangle$ and hence of L which clearly leave $L \cap \mathcal{Z}$ fixed. If $\gamma \in L^{\delta(G)}$ then $\gamma \in \mathcal{Z}$ since the image of $G \longrightarrow \mathbb{Q}\langle G \rangle$ generates $\mathbb{Q}\langle G \rangle$ as a $\mathbb{Q}$–module. Hence by Galois theory $\delta(G) = \mathrm{Gal}(L/L \cap \mathcal{Z})$. We next argue that ρ is injective on S. Suppose otherwise. Then S has an element s of prime order p in the kernel of ρ. We use the notation of Lemma 16.4. Write $s = \theta(c, f)$. Now ρ is clearly injective on T_1, so $s \notin T_1$. Hence by Lemma 16.4 either $[G : C] = 12$, $p = 3$, and 3 does not divide t; or $p = 2$ and $[G : C] = 4, 24$, or 120. In all cases s is in B and hence $f \equiv 1 \pmod{t_n}$ (Lemmas 6.9, 6.11, 7.8, and 8.4). Our hypothesis says

$c \equiv 1 \pmod{m}$ and $f \equiv 1 \pmod{n/t}$. But then $f \equiv 1 \pmod{n}$, contradicting that $s = \theta(c, f)$ has order p. Hence ρ is injective on S.

Finally suppose $g \in G$ maps to r_1 and r_2 under the maps $G \longrightarrow \mathbb{Z}_k^\bullet$ of the diagram factoring through δ and ρ, respectively. Then by definition $y^{r_1} = gyg^{-1} = y^{r_2}$ and $x^{r_1 t}\langle y \rangle = g x^t g^{-1} \langle y \rangle = x^{r_2 t} \langle y \rangle$. Hence $r_1 \equiv r_2 \pmod{m}$ and $r_1 \equiv r_2 \pmod{n/t}$, so $r_1 = r_2$. That is, the diagram commutes. □

We are now ready to give the proof of Theorem 16.1.

PROOF. Let $\overline{S}$ denote the image of S under the function $\rho : \mathbb{Z}_{|C|}^\bullet \longrightarrow \mathbb{Z}_k^\bullet$ of Lemma 16.5. Consider a fifth set of isomorphism invariants:

(D5) $|G|$, $|C|$, $|\mathcal{Z}(C)C'|$, and $\overline{S}$.

We will show that the data in (D5) determine $[G]$ and that

$$(D1) \longrightarrow (D2) \longrightarrow (D3) \longrightarrow (D5) \longleftarrow (D4)$$

where "(Di) $\longrightarrow$ (Dj)" means the data in (Di) determines that in (Dj).

An inspection of Lemmas 16.2 and 16.3 shows precisely how $|C|$ and $[\mathbb{Z}_{(G)}\langle G \rangle]$ determine $[G : C]$, so (D1) $\longrightarrow$ (D2): $[G : C] = 1$ if and only if $[\mathbb{Q}\langle G \rangle : \mathbb{Q}] = \phi(|C|)$; $[G : C] = 4$ if and only if $[\mathbb{Q}\langle G \rangle : \mathbb{Q}] = 4\phi(|C|)$ and either 3 divides $|C|$ or $3 \notin \mathbb{Z}_{(G)}\langle G \rangle^\bullet$; $[G : C] = 12$ if and only if either $[\mathbb{Q}\langle G \rangle : \mathbb{Q}] = 12\phi(|C|)$ or else $[\mathbb{Q}\langle G \rangle : \mathbb{Q}] = 4\phi(|C|)$, 3 does not divide $|C|$, and $3 \in \mathbb{Z}_{(G)}\langle G \rangle^\bullet$; $[G : C] = 24$ if and only if either $[\mathbb{Q}\langle G \rangle : \mathbb{Q}] = 24\phi(|C|)$ or else $[\mathbb{Q}\langle G \rangle : \mathbb{Q}] = 8\phi(|C|)$ and either $5 \notin \mathbb{Z}_{(G)}\langle G \rangle^\bullet$ or 5 divides $|C|$; $[G : C] = 60$ if and only if $[\mathbb{Q}\langle G \rangle : \mathbb{Q}] = 8\phi(|C|)$, $5 \in \mathbb{Z}_{(G)}\langle G \rangle^\bullet$, and 5 does not divide $|C|$; $[G : C] = 120$ if and only if $[\mathbb{Q}\langle G \rangle : \mathbb{Q}] = 16\phi(|C|)$.

Now we prove (D2) $\longrightarrow$ (D3). It suffices to show how $|C|$ is determined by $[G : C]$ and $[\mathbb{Z}_{(G)}\langle G \rangle]$. The key observations are given in

Claim 1. (A) Suppose $[G : C] = 12$. Then 3 divides n if and only if 3 divides $\deg \mathbb{Q}\langle G \rangle$ or $\mathcal{Z}$ has a primitive cube root of unity.

(B) Suppose $[G : C] = 24$. Then 3 divides n if and only if some quadratic field extension of $\mathcal{Z}$ has a primitive 9th root of unity.

Proof of Claim 1. (A) If 3 does not divide n, then $\mathcal{Z}$ is isomorphic to a subfield of $\mathbb{Q}[\zeta_k]$ which has no primitive cube roots of unity (Theorem 7.11) and 3 does not divide the degree of $\mathbb{Q}\langle G \rangle$ (Lemma 16.2). Now suppose 3 divides n and 3 does not divide $\deg \mathbb{Q}\langle G \rangle$. By Theorem 7.13 and the paragraph following it, the invariant of G must be trivial and hence $\mathcal{Z}$ has a root of unity of order $3(3_n)$. Thus $\mathcal{Z}$ has a primitive cube root of unity, as required.

(B) Suppose 3 divides n. G has a unique subgroup H which is an A_4–complement with the same core as G. By Theorem 8.7C the center of $\mathbb{Q}\langle H \rangle$ is an extension of $\mathcal{Z}$ of degree at most 2. H has trivial invariant, so by Theorem 7.13 the center of $\mathbb{Q}\langle H \rangle$ has a root of unity of order $3(3_n)$ and hence a primitive 9th root of unity. Now suppose 3 does not divide n. Again by Theorem 8.7C, $\mathcal{Z}$ is isomorphic to a subfield of $E[\sqrt{2}]$, where E is the center of $\mathbb{Q}\langle H \rangle$. By Theorem 7.11, $\mathcal{Z}$ is then isomorphic to a subfield of $\mathbb{Q}[\zeta_k, \zeta_8] = \mathbb{Q}[\zeta_{4k}]$ since $\zeta_8 + \zeta_8^{-1} = \sqrt{2}$. Now $\mathbb{Q}[\zeta_9, \zeta_{4k}]$ is an extension of $\mathbb{Q}[\zeta_{4k}]$ of degree 6, and hence no subfield of $\mathbb{Q}[\zeta_{4k}]$ has a quadratic extension containing ζ_9. Thus $\mathcal{Z}$ has no quadratic extension with a primitive 9th root of unity. This completes the proof of Claim 1.

We now apply the identity
$$|C| = \phi(|C|)|C|_0/\phi(|C|_0)$$
(cf. formula (24) in Chapter 15) to the computation of $|C|$ in terms of $[G : C]$ and $[\mathbb{Z}_{(G)}\langle G\rangle]$. By Lemma 16.3 $|G|_0$ is the product of the distinct rational primes in $\mathbb{Z}_{(G)}\langle G\rangle^\bullet$. Then a routine computation using Claim 1 above and Lemma 16.2 shows that
$$|C| = |G|_0[\mathbb{Q}\langle G\rangle : \mathbb{Q}]/(\phi(|G|_0)\delta)$$
where $\delta = 1$ if $[G : C] = 1$; $\delta = 4$ if $[G : C] = 4$; $\delta = 12$ if $[G : C] = 12$ and either 3 divides $\deg \mathbb{Q}\langle G\rangle$ or $\mathcal{Z}$ has a primitive cube root of unity; $\delta = 6$ if $[G : C] = 12$ and 3 does not divide $\deg \mathbb{Q}\langle G\rangle$ and $\mathcal{Z}$ has no primitive cube root of unity; $\delta = 12$ if $[G : C] = 24$ and no quadratic extension of $\mathcal{Z}$ has a primitive 9th root of unity; $\delta = 24$ if $[G : C] = 24$ and some quadratic extension of $\mathcal{Z}$ has a primitive 9th root of unity; $\delta = 15$ if $[G : C] = 60$; and $\delta = 30$ if $[G : C] = 120$.

Our next task is to show (D3) $\longrightarrow$ (D5). We need the following technical result.

Claim 2. The degree of $\mathbb{Q}\langle G\rangle$ is $6t$ if and only if $[G : C] = 12$, 3 divides $|C|$, and $\mathcal{Z}$ does not have a root of unity of order $3(3_{|C|})$.

Proof of Claim 2. By Lemma 16.2 we may assume without loss of generality that $[G : C] = 12$ and 3 divides $|C|$. Then by Theorem 7.13 the degree is $6t$ if and only if the reduced invariant of G is nontrivial. In this case $\mathcal{Z}$ is isomorphic to a subfield of $\mathbb{Q}[\zeta_k]$, which clearly has no root of unity of order $3(3_n)$. On the other hand if the reduced invariant is trivial, $\mathcal{Z}$ is isomorphic to $K[z(1 - u - v - vu)]$ (in the notation of Theorem 7.13) which has $-z(1 - u - v - vu)/2$ as a root of unity of order $3(3_n)$. This completes the proof of Claim 2.

We now show how $k = |\mathcal{Z}(C)C'|$ is determined by $|G|$, $|C|$, and $[\mathcal{Z}]$. First suppose $|G| = |C|$. Then by Lemma 16.2 and the definition of degree, the square of the degree of $\mathbb{Q}\langle G\rangle$ is
$$t^2 = [\mathbb{Q}\langle G\rangle : \mathbb{Q}][\mathcal{Z} : \mathbb{Q}]^{-1} = \phi(|C|)/[\mathcal{Z} : \mathbb{Q}],$$
so that
$$|\mathcal{Z}(C)C'| = |C|/(\phi(|C|)/[\mathcal{Z} : \mathbb{Q}])^{\frac{1}{2}}.$$
The other cases are similar but a bit more complex. Set $\xi = 3$ if $|G| = 12|C|$, 3 divides $|C|$, and $\mathcal{Z}$ has no root of unity of order $3(3_{|C|})$, and let $\xi = 1$ otherwise. Since $|G| \neq |C|$, then t is odd and hence by Claim 2 and Lemma 16.2, $t^2 = \xi^{-2}(\deg \mathbb{Q}\langle G\rangle)^2//2$, which by the definition of degree and Lemma 16.2 equals
$$\xi^{-2}[\mathbb{Q}\langle G\rangle : \mathbb{Q}][\mathcal{Z} : \mathbb{Q}]^{-1}//2$$
$$= \xi^{-2}(\phi(|C|)/[\mathcal{Z} : \mathbb{Q}])([\mathbb{Q}\langle G\rangle : \mathbb{Q}]/\phi(|C|))//2$$
$$= \xi^{-2}(3, |C|, |G|/|C|)(\phi(|C|)/[\mathcal{Z} : \mathbb{Q}])//2.$$
This shows t, and hence $k = |\mathcal{Z}(C)C'| = |C|/t$, are both determined by $|C|$, $|G|$, and $[\mathcal{Z}]$.

It remains to show that $\overline{S}$ is determined by $|G|$, $|C|$, and $[\mathcal{Z}]$.

By Theorem 12.6 (and Remark 12.7) $\mathbb{Q}\langle G\rangle$ has a subfield W containing $\mathcal{Z}$ and L which is isomorphic to a subfield of $U := \mathbb{Q}[\zeta_{60k}]$. By Galois theory each subfield V of W is isomorphic to only one subfield, call it V_1, of U. Thus for example L_1 is $\mathbb{Q}[\zeta_k]$, the unique subfield of U isomorphic to L. Since any homomorphism $W \longrightarrow U$

preserves intersection of subfields of W, then $L_1 \cap \mathcal{Z}_1$ is the unique subfield of U isomorphic to $L \cap \mathcal{Z}$. Thus the image of the natural map

$$\text{Gal}(L_1/L_1 \cap \mathcal{Z}_1) \longrightarrow \mathbb{Z}_k^\bullet$$

depends only on $[\mathcal{Z}]$ and $k = |\mathcal{Z}(C)C'|$ and it equals the image of

$$\text{Gal}(L/L \cap \mathcal{Z}) \longrightarrow \mathbb{Z}_k^\bullet$$

which by Lemma 16.5 is exactly the image of S in $|\mathbb{Z}_k^\bullet|$. This completes the proof that (D3) $\longrightarrow$ (D5).

The penultimate step in the proof of Theorem 16.1 is to show that (D4) $\longrightarrow$ (D5). This amounts to showing how $k = |\mathcal{Z}(C)C'|$ is determined by $|G|$, $|C|$ and S. If $|G| = |C|$ then $k = mn/t = |C|/|S|$. Now suppose $|G| \neq |C|$. If $|G|/|C| \neq 12$ or 3 does not divide $|S|$, then by Lemma 16.4 $t = |S|//2$, so

$$k = |C|/(|S|//2).$$

Finally suppose $|G|/|C| = 12$ and 3 divides $|S|$. Then by Lemma 16.4 $t = |S|/3$, so $k = 3|C|/|S|$.

It remains to show that $[G]$ is determined by the data in (D5), i.e., by $|G|$, $|C|$, $k = |\mathcal{Z}(C)C'|$ and $\overline{S}$. With δ as in Lemma 16.5 we clearly have $\langle \overline{x}t\overline{y}\rangle^{\delta(C)} \supset \langle \overline{x}t\rangle$; indeed we have equality since if for some $s \in \mathbb{Z}$ $(yx^t)^s = x(yx^t)^s x^{-1}$, then $y^s x^{ts} = y^{rs} x^{ts}$, so $s(r-1) \equiv 0 \pmod{m}$. Then $s \equiv 0 \pmod{m}$ (recall that $(r-1, m) = 1$), so $(yx^t)^s \in \langle x^t \rangle$. For any integer j let $\overline{S}_j$ denote the set of members of $\overline{S}$ of order dividing j. Note $t = |C|/k$ and hence $\overline{S}_t$ is determined by the data in (D5). By Lemmas 16.4 and 16.5, $\overline{S}_t$ is the image of $\delta(C)$ in $\mathbb{Z}_k^\bullet$ under the canonical isomorphism $\text{Aut } L \longrightarrow \mathbb{Z}_k^\bullet$ and hence

$$\left|\{\gamma \in \langle \zeta_k \rangle : \gamma^s = \gamma \text{ for all } s \in \overline{S}_t\}\right|$$
$$= \left|\langle \overline{x}t\overline{y}\rangle^{\delta(C)}\right| = \left|\langle \overline{x}t\rangle\right| = n/t$$

is determined by the data in (D5), as are $n = (n/t)t$, $m = |C|/n$, and

$$T = \text{image of } \overline{S}_t \text{ under the map } \mathbb{Z}_k^\bullet \longrightarrow \mathbb{Z}_m^\bullet.$$

If $|G|/|C| = 1$ or 60, this says $[G]$ is determined by the data in (D5) (cf. Theorem 5.2 and 9.1). The same conclusion can be drawn for all other possible values of $|G|/|C|$ since if $|G|/|C|$ is 4, 12, 24, or 120, then the reduced invariant of G is precisely the image of $\overline{S}_2$, $\overline{S}_{3(3_n)}$, $\overline{S}_2$, or $\overline{S}_2$ under the canonical maps from $\mathbb{Z}_k^\bullet$ into $\mathbb{Z}_m^\bullet \times \mathbb{Z}_{n//t}^\bullet$, $\mathbb{Z}_m^\bullet \times \mathbb{Z}_{n//6t}^\bullet$, $\mathbb{Z}_m^\bullet \times \mathbb{Z}_{n//6t}^\bullet$, or $\mathbb{Z}_m^\bullet \times \mathbb{Z}_{n//6t}^\bullet$, respectively (cf. Lemma 16.4). We are using Theorems 6.5, 7.3, 8.2, and 10.2, which say that such a Frobenius complement is determined up to isomorphism by its core index, core invariant, and reduced invariant. This completes the proof of Theorem 16.1. □

CHAPTER 17

Schur Indices and Finite Subgroups of Division Rings

A formula is established here for the Schur index of the rational truncated group ring of a 1-complement in terms of its invariant. The next two propositions show that the problem of computing the degree and index of $\mathbb{Q}\langle G\rangle$ for any Frobenius complement G can be regarded as an extension of the problem of finding all finite subgroups of division rings (*i.e.*, finite subgroups of the groups of multiplicative units of division rings), since the finite subgroups of division rings are exactly the Frobenius complements whose rational truncated group rings have index and degree equal. (Recall that a central simple algebra is a division ring if and only if its degree equals its index.)

17.1. PROPOSITION. [**SW**, Theorem 2.1.2, p. 45]. *A nontrivial finite subgroup of a division ring is a Frobenius complement.*

PROOF. Let D be a division ring and H be a nontrivial finite subgroup of $D^\bullet$. Suppose D has finite characteristic p; then the image of the canonical unitary homomorphism $\mathbb{Z}H \longrightarrow D$ is a finite ring without zero divisors, and hence is a finite field by Wedderburn's Theorem on finite division rings. Thus H is cyclic (this is essentially the argument of [**H1**, Theorem 6, p. 122]). It follows from Corollary 3.3 that H is a Frobenius complement. Now suppose D has characteristic zero. By Lemma 2.4 we have a unitary homomorphism $\mathbb{Z}\langle H\rangle \longrightarrow D$, so $\mathbb{Z}\langle H\rangle$ cannot be a torsion $\mathbb{Z}$–module. Hence H is a Frobenius complement (Lemma 2.2). □

17.2. PROPOSITION. *(Cf.* [**Am**, Theorem 3, p. 364.].*) A Frobenius complement G is a finite subgroup of a division ring if and only if $\mathbb{Q}\langle G\rangle$ is a division ring.*

PROOF. Sufficiency follows from Corollary 12.3. Now suppose G is a subgroup of a division ring D. Without loss of generality we may suppose G is not cyclic (cf. Theorem 3.2), so D has characteristic zero (argue as in the preceeding proof or apply [**H1**, Theorem 6, p. 122]). Then the canonical homomorphism $\mathbb{Z}\langle G\rangle \longrightarrow D$ (Lemma 2.4) extends to a unitary homomorphism $\mathbb{Q}\langle G\rangle \longrightarrow D$, which is injective since $\mathbb{Q}\langle G\rangle$ is simple. Therefore $\mathbb{Q}\langle G\rangle$ has no zero divisors, and hence by Wedderburn's theorem on simple algebras it must be a division ring. □

The above propositions suggest that rational truncated group rings can provide a convenient setting for studying finite subgroups of division rings. We use them now in a proof (with no claim of originality) of Amitsur's theorem [**Am**, Corollary 4, p. 384] that there is only one nonsolvable finite subgroup of a division ring.

17.3. THEOREM. *H_{120} is the only (up to isomorphism) nonsolvable finite subgroup of a division ring.*

PROOF. Suppose H is an nonsolvable finite subgroup of a division ring. By Theorem 17.1 H is a Frobenius complement and hence without loss of generality H has a subgroup of index $s \leq 2$ of the form $H_{120} \times J$ where J is a $\mathbb{Z}$–group of order relatively prime to 30 [**Pa**, Theorem 18.6, p. 204]. It suffices to show s and J are trivial.

First suppose $J \neq 1$, so it is also a Frobenius complement. Suppose J has invariant $(m, n, \langle r \rangle)$ and x, y is an r–sequence for J. Then $J_0 := H_{120} \times \langle x \rangle$ is a subgroup of H; let

$$\mathcal{C} := \mathbb{Q}\langle J_0 \rangle \cong \mathbb{Q}[\mathbf{i}, \mathbf{j}, \sqrt{5}\,] \otimes \mathbb{Q}[\zeta_n]$$

(cf. the Direct Product Lemma 2.9 and Theorem 3.2). By Theorem 17.2, $\mathcal{C}$ is a division ring. Now $\mathcal{C}$ is clearly the quaternion algebra $\left(\frac{-1,-1}{E}\right)$ over the field $E := \mathbb{Q}[\sqrt{5}, \zeta_n]$ (note that 5 does not divide n, so $\sqrt{5} \notin \mathbb{Q}[\zeta_n]$). Since $\mathcal{C}$ is a division ring, the quadratic form $(-1)X^2 + (-1)Y^2$ does not represent 1 over E [**L**, Theorem 2.7, p. 58]. Thus the field E has level 4, so the residue class degree f for the prime 2 in the extension $E/\mathbb{Q}$ is odd [**L**, Proposition 2.11, p. 307]. (Since $n > 5$, E is not formally real.) But this is impossible, since the residue class degree for the prime 2 in the extension $\mathbb{Q}[\sqrt{5}\,]/\mathbb{Q}$ is clearly 2 (after all, $\mathbb{Q}[\sqrt{5}\,]$ is the splitting field of $x^2 + 3x + 1$ and a root of this polynomial induces a quadratic extension of $\mathbb{Z}_2$). Hence J must be trivial.

Just suppose $s = 2$. Then H may be identified with H_{240} (cf. Example 4.3 and Theorem 10.2), so $\mathbb{Q}\langle H_{240} \rangle$ has no zero divisors (Theorem 17.2). But this is false since

$$\left(1 - \gamma\widehat{\psi}\right)\left(1 + \gamma\widehat{\psi}\right) = 1 - \gamma\left(\widehat{\psi}\gamma\widehat{\psi}^{-1}\right)\mathbf{i} = 1 - \gamma\psi(\gamma)\mathbf{i} = 0$$

where $\gamma = ((2 + \sqrt{5}\,)/5)(3\mathbf{j} + 4\mathbf{k})$. Hence $s = 1$, as claimed. □

In the next theorem the computation of the index of the rational truncated group ring $\mathbb{Q}\langle G \rangle$ of a 1-complement G is reduced to tractable calculations in elementary number theory. We use the notation of Notation 6.2 for the remainder of this chapter.

17.4. THEOREM. *Let G be a 1-complement with invariant $(m, n, \langle r \rangle)$. For each prime divisor p of m let*

$$f(p) = LCM\big[|p + n/t\mathbb{Z}|,\ |\langle p + m//p\mathbb{Z}, r + m//p\mathbb{Z}\rangle|/|r + m//p\mathbb{Z}|\big]$$

and let

$$\delta = GCD\{|r + m//p\mathbb{Z}|(p^{f(p)} - 1) : p \mid m\}.$$

Then $\mathbb{Q}\langle G \rangle$ has index 1 if $m = 1$; index 2 if $m \neq 1$, $n/t = 2$ and $-1 \in \langle r \rangle$; and index $n/(n, \delta)$ otherwise.

In the statement of the above theorem all subgroups are multiplicative subgroups and all orders of elements are multiplicative, not additive, orders. Before proving the above theorem we consider some applications.

17.5. REMARK. Suppose G is a 1–complement and k is the degree of $\mathbb{Q}\langle G \rangle$ divided by its index. Then $\mathbb{Q}\langle G \rangle$ is isomorphic to a ring of $k \times k$–matrices over a division ring, so G is isomorphic to a subgroup of the group of invertible elements of such a ring. Thus k measures how close G is to being a finite subgroup of a division ring; k equals 1 if and only if G is a finite subgroup of a division ring (cf.

Theorem 17.2). We will call k the *reduced degree* of G. Here is a table giving, subject to restrictions on the *order* of the Frobenius complements, the number of *isomorphism classes* of 1–complements, the number of these which are finite *subgroups* of division rings, the *average* of their reduced degrees (rounded to two decimal places), and the *maximum* of their reduced degrees. (The words emphasized in the previous sentence correspond to the labels in the table below.)

order	isomorphism classes	subgroups	average	maximum
$\leq 10{,}000$	20,509	13,197	1.60	16
$\leq 50{,}000$	119,868	66,825	1.85	32
$\leq 100{,}000$	256,349	134,161	1.98	64

The numbers of isomorphism classes were computed using Proposition 11.1; the computations of the reduced degrees were made using the previous theorem and Theorem 5.2D.

We next apply Theorem 17.4 to prove a variant of Shirvani's characterization of the $\mathbb{Z}$–groups which are finite subgroups of division rings [**SW**, Theorem 2.1.5, p. 47]. Among such groups are the cyclic groups and the binary dihedral groups D_{4m} where m is odd. We now find all the others. By Proposition 17.1 we can without loss of generality focus our attention on 1-complements.

17.6. THEOREM. *Suppose G is a 1-complement with invariant $(m, n, \langle r \rangle)$ which is neither a cyclic nor a binary dihedral group. Then G is a finite subgroup of a division ring if and only if for all prime divisors p of m and q of $|r + p\mathbb{Z}|$, q does not divide $|r + m//p\mathbb{Z}|$, and $|p + qq_{n/t}\mathbb{Z}|$ is larger than $(\frac{q_{n/t}}{q}, \frac{p+1}{2}, 2)$ and does not divide $|p + \rho\mathbb{Z}|$ for any rational prime ρ which divides mn but not $p|r + p\mathbb{Z}|$.*

Consider an example. For all positive integers i and j one can check that there is a unique Frobenius triple of the form $(7^i, 3^{j+1}, T)$; it has $T = \langle 2^{7^i} + 7^i\mathbb{Z} \rangle$ and $|T| = 3$. Since $7 \not\equiv 1 \pmod{3 \cdot 3^j}$, then the 1-complement with invariant $(7^i, 3^{j+1}, \langle 2^{7^i} + 7^i\mathbb{Z} \rangle)$ is a finite subgroup of a division ring. Not only is the group of this type of order 63 the unique noncyclic subgroup of a division ring of minimal odd order (which gives Amitsur's answer [**Am**, Theorem 6, p.374] to a question of Herstein [**H1**]), it is the unique noncyclic Frobenius complement of minimal odd order.

The proof of Theorem 17.6 gives more compact (but less elementary) criteria for a 1-complement to be a finite subgroup of a division ring than that above (e.g., see formula (26)). In Remark 17.7 we discuss further the divisibility conditions of the theorem above and their connections with those of [**SW**, Theorem 2.1.5(c), p. 47].

PROOF. The symbol p will always denote a prime factor of m, and q and ρ will always denote primes. We use the notation and results of Theorem 17.4 and Proposition 11.1A. In particular, $t = |r|$ is the least common multiple of the $t(p) := |r + p\mathbb{Z}|$. We will several times use implicitly below the observation that if q divides $t(p)$, then $\mathbb{Z}_p^\bullet$ has an element of order q, so q divides $p - 1$ and hence $|p + q^s\mathbb{Z}|$ is a power of q for all positive integers s.

First suppose $n/t = 2$ and $-1 \in \langle r \rangle$. Then $\mathbb{Q}\langle G \rangle$ has index 2. Since G is not binary dihedral, the degree t of $\mathbb{Q}\langle G \rangle$ is not 2, so G is not a finite subgroup of a division ring (Proposition 17.2). Now t is a power of 2 (since n_0 divides n/t), so 4 divides t and hence there exists some p with 4 dividing $p-1$. The validity of the theorem follows in this case since for any q dividing $t(p)$ we have $qq_{n/t} = 4$, so $|p + qq_{n/t}\mathbb{Z}| = 1 \le (\frac{q_{n/t}}{q}, \frac{p+1}{2}, 2)$.

Now suppose that either $n/t \ne 2$ or $-1 \notin \langle r \rangle$, so the index of $\mathbb{Q}\langle G \rangle$ is $n/(n,\delta)$. Then G is a finite subgroup of a division ring if and only if $t = n/(n,\delta)$, and hence if and only if $(n,\delta) = n/t$. Now n/t divides (n,δ) since for every p it divides $p^{|p+n/t\mathbb{Z}|} - 1$. Moreover any prime power dividing (n,δ) which is relatively prime to t divides n/t. Hence G is a finite subgroup of a division ring if and only if for all prime divisors q of t, $qq_{n/t} \nmid \delta$, i.e., there exits some p with

(26) $$qq_{n/t} \nmid |r + m//p\mathbb{Z}|(p^{f(p)} - 1).$$

Since $q_{n/t}$ divides $p^{f(p)} - 1$, the condition (26) is equivalent to the condition that $q \nmid |r + m//p\mathbb{Z}|$ and $qq_{n/t} \nmid p^{f(p)} - 1$. The relation $q \nmid |r + m//p\mathbb{Z}|$ implies that p is in fact the unique prime divisor of m with $t(p)$ divisible by q. Hence G is a finite subgroup of a division ring if and only if for all p and for all divisors q of $t(p)$ we have $q \nmid |r + m//p\mathbb{Z}|$ and $|p + q^{d+1}\mathbb{Z}| \nmid f(p)$ where we have set $q^d = q_{n/t}$; or equivalently, $q \nmid |r + m//p\mathbb{Z}|$ and

(27) $$|p + q^{d+1}\mathbb{Z}| \nmid LCM[|p + n/t\mathbb{Z}|, |p + m//p\mathbb{Z}|] = |p + nm/tp_m\mathbb{Z}|.$$

Because $|p + q^{d+1}\mathbb{Z}|$ is a prime power, the assertion (27) is valid if and only if $|p + q^{d+1}\mathbb{Z}| \nmid |p + \rho_{mn/t}\mathbb{Z}|$ for every ρ dividing $mn//p$. Now for each such ρ, $|p + \rho_{mn/t}\mathbb{Z}|$ is the product of $|p + \rho\mathbb{Z}|$ and a power of ρ. Hence the condition (27) is equivalent to saying that $|p + q^{d+1}\mathbb{Z}| \nmid |p + \rho\mathbb{Z}|$ whenever ρ divides mn but not $pt(p)$ and that

(28) $$|p + q^{d+1}\mathbb{Z}| \nmid |p + q^d\mathbb{Z}|.$$

It remains to prove that the condition (28) is equivalent to

(29) $$|p + q^{d+1}\mathbb{Z}| > c \quad \text{where} \quad c = (q^{d-1}, \frac{p+1}{2}, 2).$$

Assume condition (28). Then (29) follows trivially if $c = 1$, so suppose $c \ne 1$. Then $p \equiv 3 \pmod{4}$ and 4 divides q^d. Therefore $|p + q^d\mathbb{Z}|$ is at least 2, so $|p + q^{d+1}\mathbb{Z}|$ is at least 4, which is larger than c. Now suppose the condition (29) is valid. Condition (28) is obvious if $q^d = 2$ and $p \equiv 3 \pmod 4$, so suppose otherwise. Let a be maximal with q^a dividing $p^c - 1$. By the choice of c, either q is odd or $a \ge 2$. An easy induction argument now shows that for all $i \ge 0$ we can find an integer s_i not divisible by q such that $p^{cq^i} = 1 + q^{a+i}s_i$. By our hypothesis (29), $a < d + 1$. Since $p^{cq^{d-a}} = 1 + q^d s_{d-a}$ and $p^{cq^{d+1-a}} = 1 + q^{d+1} s_{d+1-a}$, therefore

$$|p^c + q^{d+1}\mathbb{Z}| = q^{d+1-a} > |p^c + q^d\mathbb{Z}|.$$

The validity of the assertion (28) follows readily (note that $c|p^c + q^e\mathbb{Z}| = |p + q^e\mathbb{Z}|$ whenever $d \le e \in \mathbb{Z}$). $\square$

17.7. REMARK. We sketch the connections between the divisibility conditions of the above theorem and those of Shirvani's characterization of these groups [**SW**, Theorem 2.1.5(c), p. 47]. Proofs are routine and are left to the reader.

(A) The condition that q does not divide $|r + m//p\mathbb{Z}|$ for all p and q as in the above theorem is equivalent to the condition that the numbers $|r + p\mathbb{Z}|$ as p ranges over the prime divisors of m are pairwise relatively prime. This is equivalent to the condition that G has the kind of decomposition $G_0 \times \cdots G_s$ found in [**SW**, Theorem 2.1.5(c), p. 47].

(B) The condition that $|p + qq_{n/t}\mathbb{Z}| > (q_{n/t}/q, (p+1)/2, 2)$ corresponds to the conditions on "α" in [**SW**, Theorem 2.1.5(c), p. 47].

(C) The assertion that $|p + qq_{n/t}\mathbb{Z}|$ does not divide $|p + \rho\mathbb{Z}|$ for all prime divisors ρ of mn which do not divide $pt(p)$ is equivalent (in the presence of the conditions of (A) and (B) above, which imply that $|p + qq_{n/t}\mathbb{Z}| = q|p + q_{n/t}\mathbb{Z}|$) to the divisibility condition in the last sentence of [**SW**, Theorem 2.1.5(c), p. 47].

The remainder of this chapter is devoted to the proof of Theorem 17.4. We identify $\mathbb{Q}\langle G \rangle$ with the crossed product algebra $(L/K, \Phi)$ where $L = \mathbb{Q}[\zeta_{mn/t}]$; $K = L^\sigma$ where $\sigma \in \operatorname{Aut} L$ fixes $\zeta_{n/t}$ and maps ζ_m to ζ_m^r; and Φ is the factor set on $\langle \sigma \rangle$ with $\Phi(\sigma^i, \sigma^j)$ equal to $\zeta_{n/t}$ if $i + j \geq t$ and equal to 1 otherwise, where i and j range over the nonnegative integers less than t (cf. Remark 12.9 and Case 1 of the proof of Theorem 12.6). The index of $\mathbb{Q}\langle G \rangle$ is the least common multiple of the local indices, i.e., the indices of the rings $K_\mathfrak{p} \otimes_K \mathbb{Q}\langle G \rangle$ where $\mathfrak{p}$ ranges over the primes of K [**Re**, Theorem 32.19, p. 280]. Hence the index of $\mathbb{Q}\langle G \rangle$ is the least common multiple of the indices of the crossed product algebras

$$A_\mathfrak{p} := (L_\mathfrak{p}/K_\mathfrak{p}, \Phi_\mathfrak{p}),$$

where $\mathfrak{p}$ ranges over the primes of L, $L_\mathfrak{p}$ denotes the completion of L at $\mathfrak{p}$, $K_\mathfrak{p}$ denotes the closure of K in $L_\mathfrak{p}$, and $\Phi_\mathfrak{p}$ denotes the restriction of Φ to the Galois group of $L_\mathfrak{p}/K_\mathfrak{p}$, which can be identified with a subgroup of $\langle \sigma \rangle$ [**Re**, Theorem 29.13, p. 248]. We begin by considering an infinite prime $\mathfrak{p}$ of L. If $\mathfrak{p}$ is a real prime (so $n = 2$ and $L = \mathbb{Q}$) or else restricts to a complex prime of K, then $A_\mathfrak{p}$ has index 1. Now suppose it is complex but restricts to a real prime of K. Then $n/t = 2$ (otherwise K, which contains $\zeta_{n/t}$, could not have a real prime). Let $F = K_\mathfrak{p} \cap L$. Since $[L_\mathfrak{p} : K_\mathfrak{p}] = [\mathbb{C} : \mathbb{R}] = 2$, then $[L : F] = 2$. But L/K is a cyclic extension. Hence $L^{\sigma^{t/2}} = F$ and $\sigma^{t/2}$ extends uniquely to an automorphism τ of $L_\mathfrak{p}/K_\mathfrak{p}$. The only nontrivial automorphism of $\mathbb{C}/\mathbb{R}$ is complex conjugation, which takes any m-th root of unity to its inverse. Hence $\zeta_m^{-1} = \sigma^{t/2}(\zeta_m) = \zeta_m^{r^{t/2}}$. Thus $r^{t/2} = -1$. In $A_\mathfrak{p}$ we have $u(\tau)^2 = \Phi(\sigma^{t/2}, \sigma^{t/2})u(1) = \zeta_{n/t} = -1$, and $u(\tau)\gamma u(\tau)^{-1} = \tau(\gamma)$ is the complex conjugate of γ (defined with respect to $\mathfrak{p}$) for any $\gamma \in L_\mathfrak{p}$. (We are using here the crossed product notation introduced in the paragraph preceding Theorem 12.8.) Thus $A_\mathfrak{p} \cong \mathbb{H}$ and hence $A_\mathfrak{p}$ has index 2. We end this discussion of real primes of K by showing that if $m \neq 1$, $n/t = 2$, and $r^{t/2} = -1$, then K does indeed have a real prime. After all we then have $\sigma^{t/2}(\zeta_m) = \zeta_m^{-1}$, so $\sigma^{t/2}$ is complex conjugation. Thus $K = L^\sigma \subset L^{\sigma^{t/2}} \subset \mathbb{R}$, so K has a real prime.

We now consider a finite prime $\mathfrak{p}$ of L. $\mathfrak{p}$ extends a rational prime, say p. Let us denote the residue class degree and ramification index of (the restriction of) $\mathfrak{p}$ with respect to a field extension E_1/E_2, where $L \supset E_1 \supset E_2 \supset \mathbb{Q}$, by $f(E_1/E_2)$ and $e(E_1/E_2)$, respectively. If p does not divide mn, then $\mathfrak{p}$ is unramified and hence $A_\mathfrak{p}$ has index 1 [**Jz**, p. 699]. Now suppose p divides n. By [**R**, 4B(5), p. 269]

$$e(L/\mathbb{Q}) = \phi(p_{mn/t}) = \phi(p_{n/t}) = e(\mathbb{Q}[\zeta_{n/t}]/\mathbb{Q}).$$

Hence $e(L/\mathbb{Q}[\zeta_{n/t}]) = 1$, so $e(L/K) = 1$. Thus again L/K is unramified, so $A_\mathfrak{p}$ has index 1. Finally suppose $p \mid m$. Then p does not divide $t = [L : K]$, so the extension L/K is tamely ramified. Since the automorphism group $\langle \sigma \rangle$ of L/K is cyclic, then $\theta := \sigma^{t/e(L/K)}$ must be a generator of the inertia group of L/K with respect to $\mathfrak{p}$.

Let φ be a Frobenius automorphism for L/K. Observe that $u(\theta)^{e(L/K)} = \zeta_{n/t}$ and

$$u(\varphi)u(\theta)u(\varphi)^{-1}u(\theta)^{-q} = (u(\theta)^{e(L/K)})^{-(q-1)/e(L/K)}$$

where $q = p^{f(K/\mathbb{Q})}$ since all the $u(\sigma^i)$ ($0 \leq i < t$) commute. (Note that $e(L/K)$ divides $q-1$ since it divides $([L:K], e(L/\mathbb{Q})) = (t, \phi(p_m)) = (t, p-1)$.) Since $L_\mathfrak{p}/K_\mathfrak{p}$ is tamely ramified and by definition $\Phi_\mathfrak{p}$ takes its values in the group of units of the valuation ring of $L_\mathfrak{p}$, then by [**Jz**, Theorem 2, p. 700] the index of $A_\mathfrak{p}$ is the order of the image in the residue class field $\bar{L}$ of $L_\mathfrak{p}$ of the multiplicative group

$$\langle \zeta_{n/t}^{(q-1)/e(L/K)}, \zeta_{n/t}^{(q^{f(L/K)}-1)/e(L/K)} \rangle = \langle \zeta_{n/t}^{(q-1)/e(L/K)} \rangle = \langle \zeta_s \rangle$$

where $s = (n/t)/(n/t, (q-1)/e(L/K))$. The factorization of $X^s - 1$ in L induces the factorization of $X^s - 1$ in $\bar{L}$. But because the characteristic p of $\bar{L}$ does not divide s, $X^s - 1$ factors into distinct linear factors in $\bar{L}$. Therefore the group $\langle \zeta_s \rangle$ injects into $\bar{L}$, so the index of $A_\mathfrak{p}$ is

$$(30) \qquad \frac{n/t}{(n/t, (q-1)/e(L/K))} = \frac{n}{(n, t(q-1)/e(L/K))}.$$

Next we turn to the calculation of $q = p^{f(K/\mathbb{Q})}$ and of $e(L/K)$. The ramification index $e(L/K)$ is the order of the inertia group $I(L/K)$ of $\mathfrak{p}$ with respect to the extension L/K, and this group is the intersection of the inertia group $I(L/\mathbb{Q})$ of $\mathfrak{p}$ with respect to $L/\mathbb{Q}$ and the Galois group $\langle \sigma \rangle$. We will treat as identifications the natural isomorphisms

$$(31) \qquad \begin{array}{ccc} \operatorname{Aut} \mathbb{Q}[\zeta_{mn/t}] & \longrightarrow & \operatorname{Aut} \mathbb{Q}[\zeta_{p_m}] \times \operatorname{Aut} \mathbb{Q}[\zeta_{mn/tp_m}] \\ \downarrow & & \downarrow \\ \mathbb{Z}_{mn/t}^\bullet & \longrightarrow & \mathbb{Z}_{p_m}^\bullet \times \mathbb{Z}_{mn/tp_m}^\bullet \end{array}$$

(the top map is induced by restriction). The inertia and decomposition groups of $\mathfrak{p}$ with respect to the extension $L/\mathbb{Q}$ map under the top isomorphism of the commutative diagram (31) into the direct product of the corresponding groups with respect to the extensions $\mathbb{Q}[\zeta_{p_m}]/\mathbb{Q}$ and $\mathbb{Q}[\zeta_{mn/tp_m}]/\mathbb{Q}$. We can pick $s \in \mathbb{Z}$ with $s \equiv r \pmod{m}$ and $s \equiv 1 \pmod{n/t}$. Then $\langle \sigma \rangle$ is identified with $\langle (s + p_m\mathbb{Z}, s + mn/tp_m\mathbb{Z}) \rangle$ and $I(L/\mathbb{Q})$ with $\mathbb{Z}_{p_m}^\bullet \times 1$ [**R**, 4B(5), p. 269], so $|I(L/\mathbb{Q})| = \phi(p_m)$. The intersection of these two groups is generated by $\sigma^{|s+mn/tp_m\mathbb{Z}|}$, so

$$e(L/K) = t/(t, |s + mn/tp_m\mathbb{Z}|).$$

Since $|s + mn/tp_m\mathbb{Z}| = |r + m//p\mathbb{Z}|$ and

$$t = |r| = LCM[|r + p_m\mathbb{Z}|, |r + m//p\mathbb{Z}|],$$

then

$$e(L/K) = |I(L/K)| = |r + p_m\mathbb{Z}|/(|r + p_m\mathbb{Z}|, |r + m//p\mathbb{Z}|)$$

and

$$(32) \qquad t/e(L/K) = |r + m//p\mathbb{Z}|.$$

Now let $D(L/\mathbb{Q})$ and $D(L/K)$ denote the decomposition groups of $\mathfrak{p}$ with respect to the indicated field extensions. Then $D(L/\mathbb{Q})$ corresponds under our identification

(31) to $\mathbb{Z}_{p_m}^{\bullet} \times \langle p + mn/tp_m\mathbb{Z} \rangle$ (the decomposition group of $\mathfrak{p}$ with respect to the field extension $\mathbb{Q}[\zeta_{mn/tp_m}]/\mathbb{Q}$ is generated by the Frobenius automorphism). Hence

$$|D(L/\mathbb{Q})| = \phi(p_m)|p + mn/tp_m\mathbb{Z}|\,.$$

$D(L/K)$ is the intersection of $D(L/\mathbb{Q})$ and $\langle \sigma \rangle$, so it is generated by σ^i and has order t/i where i is the order of the coset of $s + mn/tp_m\mathbb{Z}$ in the factor group $\mathbb{Z}_{mn/tp_m}^{\bullet}/\langle p + mn/tp_m\mathbb{Z} \rangle$. Thus $i = d_{mn/tp_m}(p,s)/|p+mn/tp_m\mathbb{Z}|$. Here we are letting $d_\mu(j,k)$ denote the order of the subgroup $D_\mu(j,k) := \langle j+\mu\mathbb{Z}, k+\mu\mathbb{Z} \rangle$ of $\mathbb{Z}_\mu^{\bullet}$ for any integers j and k relatively prime to the positive integer μ. Hence

$$|D(L/K)| = t|p + mn/tp_m\mathbb{Z}|/d_{mn/tp_m}(p,s)\,.$$

Combining these ramification group calculations we obtain

$$f(K/\mathbb{Q}) = \frac{|D(L/\mathbb{Q})||I(L/K)|}{|I(L/\mathbb{Q})||D(L/K)|} = \frac{d_{mn/tp_m}(p,s)}{|r+m//p\mathbb{Z}|}\,.$$

Now let us consider the restriction to $D_{mn/tp_m}(p,s)$ of the natural homomorphism $\mathbb{Z}_{mn/tp_m}^{\bullet} \longrightarrow \mathbb{Z}_{n/t}^{\bullet}$. Since $s+mn/tp_m\mathbb{Z}$ is in the kernel, then the image is $\langle p+n/t\mathbb{Z} \rangle$ and the kernel is $D := D_{mn/tp_m}(s, p^{|p+n/t\mathbb{Z}|})$. Since the generators of D are trivial modulo n/t, it follows that

$$d_{mn/tp_m}(p,s) = |p+n/t\mathbb{Z}||D| = |p+n/t\mathbb{Z}|d_{m//p}(s, p^{|p+n/t\mathbb{Z}|})\,.$$

The order of the coset of $p+m//p\mathbb{Z}$ in the factor group $\mathbb{Z}_{m//p}^{\bullet}/\langle s+m//p\mathbb{Z}\rangle$ is $j := d_{m//p}(p,s)/|s+m//p\mathbb{Z}|$, so the order of the coset of $p^{|p+n/t\mathbb{Z}|} + m//p\mathbb{Z}$ in the factor group is

$$\frac{j}{(j, |p+n/t\mathbb{Z}|)} = \frac{d_{m//p}(s, p^{|p+n/t\mathbb{Z}|})}{|s+m//p\mathbb{Z}|}\,.$$

Hence

$$d_{m//p}(s, p^{|p+n/t\mathbb{Z}|}) = \frac{|s+m//p\mathbb{Z}|d_{m//p}(p,s)}{(d_{m//p}(p,s), |s+m//p\mathbb{Z}||p+n/t\mathbb{Z}|)}\,.$$

Because $r \equiv s \pmod{m}$ we can deduce that

$$\begin{aligned} f(K/\mathbb{Q}) &= \frac{d_{m//p}(p,s)|p+n/t\mathbb{Z}||s+m//p\mathbb{Z}|}{(d_{m//p}(p,s), |p+n/t\mathbb{Z}||s+m//p\mathbb{Z}|)|r+m//p\mathbb{Z}|} \\ &= LCM[d_{m//p}(p,s), |p+n/t\mathbb{Z}||s+m//p\mathbb{Z}|]/|r+m//p\mathbb{Z}| \\ &= f(p)\,. \end{aligned}$$

Hence by formulas (30) and (32) the index of $A_{\mathfrak{p}}$ is

$$n/\bigl(n, (p^{f(p)}-1)|r+m//p\mathbb{Z}|\bigr)\,.$$

We can now give the index of $\mathbb{Q}\langle G \rangle$. If $m = 1$, then $\mathbb{Q}\langle G \rangle \cong L$ is abelian, so its index is indeed 1. Next suppose that $m \neq 1$, $n/t = 2$, and $r^{t/2} = -1$. Then we have seen that K has a real prime and the local index at any such prime is 2. On the other hand by equation (30) the local index at any finite prime divides $n/t = 2$. Thus the index of $\mathbb{Q}\langle G \rangle$ is exactly 2. Finally suppose that neither of the above two cases occurs. Then K has no real primes, so the index of $\mathbb{Q}\langle G \rangle$ is the least common multiple of the indices of the rings $A_\mathfrak{p}$ as $\mathfrak{p}$ ranges over finite primes which extend rational primes p dividing m. That is, the index is

$$LCM\{\frac{n}{\left(n, |r + m//p\mathbb{Z}|(p^{f(p)} - 1)\right)} : p \mid m\} = \frac{n}{(n, \delta)}.$$

This completes the proof of Theorem 17.4.

Bibliography

[A] A. A. Albert, *Structure of Algebras*, Amer. Math. Soc. Colloq. Publ. **24**, Providence, RI, 1961.

[Am] S. A. Amitsur, *Finite subgroups of division rings*, Trans. Amer. Math. Soc. **80** (1955), 361–386.

[B] W. Burnside, *Note on the symmetric group*, Proc. London Math. Soc. **28** (1897), 119–129.

[BrH] R. Brown and D. K. Harrison, *Abelian Frobenius kernels and modules over number rings*, J. Pure Appl. Algebra **126** (1998), 51–86.

[C] H. S. M. Coxeter, *The binary polyhedral groups, and other generalizations of the quaternion group*, Duke Math. J. **7** (1940), 367–379.

[GW] R. Guralnick and R. Wiegand, *Galois groups and the multiplicative structure of field extensions*, Trans. Amer. Math. Soc. **331** (1992), 563-584.

[H] I. N. Herstein, *Topics in Algebra*, 2nd ed., Xerox, Lexington, MA, 1975.

[H1] _____, *Finite multiplicative subgroups in division rings*, Pacific J. Math. **3** (1953), 121–126.

[J] N. Jacobson, *The Theory of Rings*, Amer. Math. Soc. Math. Surveys **2**, Providence, RI, 1943.

[J1] _____, *Basic Algebra II*, Freeman, San Francisco, 1980.

[Jz] G. J. Janusz, *Crossed product orders and the Schur index*, Comm. Algebra **8(7)** (1980), 697–706.

[L] T. Y. Lam, *The Algebraic Theory of Quadratic Forms*, Benjamin/Cummings, London, 1973.

[LL] _____ and K. H. Leung, *On vanishing sums of roots of unity*, J. Algebra **224** (2000), 91-109.

[MR] J. C. McConnell and J. C. Robson, *Noncommutative Noetherian Rings*, John Wiley, New York, 1987.

[NZ] I. Niven and H. S. Zuckerman, *An Introduction to the Theory of Numbers*, 4th ed., Wiley, New York, 1980.

[Pa] D. Passman, *Permutation Groups*, W. A. Benjamin, New York, 1968.

[P] R. S. Pierce, *Associative Algebras*, Springer–Verlag, New York, 1982.

[Re] I. Reiner, *Maximal Orders*, Academic Press, New York, 1975.

[R] P. Ribenboim, *Algebraic Numbers*, John Wiley, New York, 1972.

[S] W. R. Scott, *Group Theory*, Prentice–Hall, Englewood Cliffs, NJ, 1964.

[SW] M. Shirvani and B. A. F. Wehrfritz, *Skew Linear Groups*, Cambridge University Press, Cambridge, 1986.

[T] J. Thompson, *Finite groups with fixed-point-free automorphisms of prime order*, Proc. Natl. Acad. Sci. U. S. A. **45** (1959), 578–581.

[V] M–F. Vignéras, *Arithmétique des Algèbres de Quaternions*, Lec. Notes in Math. 800, Springer–Verlag, Berlin, 1980.

[Z] H. Zassenhaus, *Über endliche Fastkörper*, Abh. Math. Sem. Univ. Hamburg **11** (1936), 187–220.

Editorial Information

To be published in the *Memoirs*, a paper must be correct, new, nontrivial, and significant. Further, it must be well written and of interest to a substantial number of mathematicians. Piecemeal results, such as an inconclusive step toward an unproved major theorem or a minor variation on a known result, are in general not acceptable for publication. Papers appearing in *Memoirs* are generally longer than those appearing in *Transactions*, which shares the same editorial committee.

As of January 31, 2001, the backlog for this journal was approximately 7 volumes. This estimate is the result of dividing the number of manuscripts for this journal in the Providence office that have not yet gone to the printer on the above date by the average number of monographs per volume over the previous twelve months, reduced by the number of volumes published in four months (the time necessary for preparing a volume for the printer). (There are 6 volumes per year, each containing at least 4 numbers.)

A Consent to Publish and Copyright Agreement is required before a paper will be published in the *Memoirs*. After a paper is accepted for publication, the Providence office will send a Consent to Publish and Copyright Agreement to all authors of the paper. By submitting a paper to the *Memoirs*, authors certify that the results have not been submitted to nor are they under consideration for publication by another journal, conference proceedings, or similar publication.

Information for Authors

Memoirs are printed from camera copy fully prepared by the author. This means that the finished book will look exactly like the copy submitted.

The paper must contain a *descriptive title* and an *abstract* that summarizes the article in language suitable for workers in the general field (algebra, analysis, etc.). The *descriptive title* should be short, but informative; useless or vague phrases such as "some remarks about" or "concerning" should be avoided. The *abstract* should be at least one complete sentence, and at most 300 words. Included with the footnotes to the paper should be the 2000 *Mathematics Subject Classification* representing the primary and secondary subjects of the article. The classifications are accessible from **www.ams.org/msc/**. The list of classifications is also available in print starting with the 1999 annual index of *Mathematical Reviews*. The Mathematics Subject Classification footnote may be followed by a list of *key words and phrases* describing the subject matter of the article and taken from it. Journal abbreviations used in bibliographies are listed in the latest *Mathematical Reviews* annual index. The series abbreviations are also accessible from **www.ams.org/publications/**. To help in preparing and verifying references, the AMS offers MR Lookup, a Reference Tool for Linking, at **www.ams.org/mrlookup/**. When the manuscript is submitted, authors should supply the editor with electronic addresses if available. These will be printed after the postal address at the end of the article.

Electronically prepared manuscripts. The AMS encourages electronically prepared manuscripts, with a strong preference for $\mathcal{A}_{\mathcal{M}}\mathcal{S}$-LaTeX. To this end, the Society has prepared $\mathcal{A}_{\mathcal{M}}\mathcal{S}$-LaTeX author packages for each AMS publication. Author packages include instructions for preparing electronic manuscripts, the *AMS Author Handbook*, samples, and a style file that generates the particular design specifications of that publication series. Though $\mathcal{A}_{\mathcal{M}}\mathcal{S}$-LaTeX is the highly preferred format of TeX, author packages are also available in $\mathcal{A}_{\mathcal{M}}\mathcal{S}$-TeX.

Authors may retrieve an author package from e-MATH starting from `www.ams.org/tex/` or via FTP to `ftp.ams.org` (login as `anonymous`, enter username as password, and type `cd pub/author-info`). The *AMS Author Handbook* and the *Instruction Manual* are available in PDF format following the author packages link from `www.ams.org/tex/`. The author package can be obtained free of charge by sending email to `pub@ams.org` (Internet) or from the Publication Division, American Mathematical Society, P.O. Box 6248, Providence, RI 02940-6248. When requesting an author package, please specify $\mathcal{AMS}$-LaTeX or $\mathcal{AMS}$-TeX, Macintosh or IBM (3.5) format, and the publication in which your paper will appear. Please be sure to include your complete mailing address.

Sending electronic files. After acceptance, the source file(s) should be sent to the Providence office (this includes any TeX source file, any graphics files, and the DVI or PostScript file).

Before sending the source file, be sure you have proofread your paper carefully. The files you send must be the EXACT files used to generate the proof copy that was accepted for publication. For all publications, authors are required to send a printed copy of their paper, which exactly matches the copy approved for publication, along with any graphics that will appear in the paper.

TeX files may be submitted by email, FTP, or on diskette. The DVI file(s) and PostScript files should be submitted only by FTP or on diskette unless they are encoded properly to submit through email. (DVI files are binary and PostScript files tend to be very large.)

Electronically prepared manuscripts can be sent via email to `pub-submit@ams.org` (Internet). The subject line of the message should include the publication code to identify it as a Memoir. TeX source files, DVI files, and PostScript files can be transferred over the Internet by FTP to the Internet node `e-math.ams.org` (130.44.1.100).

Electronic graphics. Comprehensive instructions on preparing graphics are available at `www.ams.org/jourhtml/graphics.html`. A few of the major requirements are given here.

Submit files for graphics as EPS (Encapsulated PostScript) files. This includes graphics originated via a graphics application as well as scanned photographs or other computer-generated images. If this is not possible, TIFF files are acceptable as long as they can be opened in Adobe Photoshop or Illustrator. No matter what method was used to produce the graphic, it is necessary to provide a paper copy to the AMS.

Authors using graphics packages for the creation of electronic art should also avoid the use of any lines thinner than 0.5 points in width. Many graphics packages allow the user to specify a "hairline" for a very thin line. Hairlines often look acceptable when proofed on a typical laser printer. However, when produced on a high-resolution laser imagesetter, hairlines become nearly invisible and will be lost entirely in the final printing process.

Screens should be set to values between 15% and 85%. Screens which fall outside of this range are too light or too dark to print correctly. Variations of screens within a graphic should be no less than 10%.

Inquiries. Any inquiries concerning a paper that has been accepted for publication should be sent directly to the Electronic Prepress Department, American Mathematical Society, P. O. Box 6248, Providence, RI 02940-6248.

Editors

This journal is designed particularly for long research papers, normally at least 80 pages in length, and groups of cognate papers in pure and applied mathematics. Papers intended for publication in the *Memoirs* should be addressed to one of the following editors. In principle the Memoirs welcomes electronic submissions, and some of the editors, those whose names appear below with an asterisk (*), have indicated that they prefer them. However, editors reserve the right to request hard copies after papers have been submitted electronically. Authors are advised to make preliminary email inquiries to editors about whether they are likely to be able to handle submissions in a particular electronic form.

Algebra to CHARLES CURTIS, Department of Mathematics, University of Oregon, Eugene, OR 97403-1222 email: cwc@darkwing.uoregon.edu

Algebraic geometry and commutative algebra to LAWRENCE EIN, Department of Mathematics, University of Illinois, 851 S. Morgan (M/C 249), Chicago, IL 60607-7045; email: ein@uic.edu

Algebraic topology and cohomology of groups to STEWART PRIDDY, Department of Mathematics, Northwestern University, 2033 Sheridan Road, Evanston, IL 60208-2730; email: priddy@math.nwu.edu

Combinatorics and Lie theory to SERGEY FOMIN, Department of Mathematics, University of Michigan, Ann Arbor, Michigan 48109-1109; email: fomin@math.lsa.umich.edu

Complex analysis and complex geometry to DUONG H. PHONG, Department of Mathematics, Columbia University, 2990 Broadway, New York, NY 10027-0029; email: dp@math.columbia.edu

***Differential geometry and global analysis** to LISA C. JEFFREY, Department of Mathematics, University of Toronto, 100 St. George St., Toronto, ON Canada M5S 3G3; email: jeffrey@math.toronto.edu

***Dynamical systems and ergodic theory** to ROBERT F. WILLIAMS, Department of Mathematics, University of Texas, Austin, Texas 78712-1082; email: bob@math.utexas.edu

Geometric topology, knot theory and hyperbolic geometry to ABIGAIL A. THOMPSON, Department of Mathematics, University of California, Davis, Davis, CA 95616-5224; email: thompson@math.ucdavis.edu

Harmonic analysis, representation theory, and Lie theory to ROBERT J. STANTON, Department of Mathematics, The Ohio State University, 231 West 18th Avenue, Columbus, OH 43210-1174; email: stanton@math.ohio-state.edu

***Logic** to THEODORE SLAMAN, Department of Mathematics, University of California, Berkeley, CA 94720-3840; email: slaman@math.berkeley.edu

Number theory to MICHAEL J. LARSEN, Department of Mathematics, Indiana University, Bloomington, IN 47405; email: larsen@math.indiana.edu

Operator algebras and functional analysis to BRUCE E. BLACKADAR, Department of Mathematics, University of Nevada, Reno, NV 89557; email: bruceb@math.unr.edu

***Ordinary differential equations, partial differential equations, and applied mathematics** to PETER W. BATES, Department of Mathematics, Brigham Young University, 292 TMCB, Provo, UT 84602-1001; email: peter@math.byu.edu

***Partial differential equations and applied mathematics** to BARBARA LEE KEYFITZ, Department of Mathematics, University of Houston, 4800 Calhoun Road, Houston, TX 77204-3476; email: keyfitz@uh.edu

***Probability and statistics** to KRZYSZTOF BURDZY, Department of Mathematics, University of Washington, Box 354350, Seattle, Washington 98195-4350; email: burdzy@math.washington.edu

***Real and harmonic analysis and geometric partial differential equations** to WILLIAM BECKNER, Department of Mathematics, University of Texas, Austin, TX 78712-1082; email: beckner@math.utexas.edu

All other communications to the editors should be addressed to the Managing Editor, WILLIAM BECKNER, Department of Mathematics, University of Texas, Austin, TX 78712-1082; email: beckner@math.utexas.edu.

Selected Titles in This Series

(Continued from the front of this publication)

687 **Guy David and Stephen Semmes,** Uniform rectifiability and quasiminimizing sets of arbitrary codimension, 2000

686 **L. Gaunce Lewis, Jr.,** Splitting theorems for certain equivariant spectra, 2000

685 **Jean-Luc Joly, Guy Metivier, and Jeffrey Rauch,** Caustics for dissipative semilinear oscillations, 2000

684 **Harvey I. Blau, Bangteng Xu, Z. Arad, E. Fisman, V. Miloslavsky, and M. Muzychuk,** Homogeneous integral table algebras of degree three: A trilogy, 2000

683 **Serge Bouc,** Non-additive exact functors and tensor induction for Mackey functors, 2000

682 **Martin Majewski,** ational homotopical models and uniqueness, 2000

681 **David P. Blecher, Paul S. Muhly, and Vern I. Paulsen,** Categories of operator modules (Morita equivalence and projective modules, 2000

680 **Joachim Zacharias,** Continuous tensor products and Arveson's spectral C^*-algebras, 2000

679 **Y. A. Abramovich and A. K. Kitover,** Inverses of disjointness preserving operators, 2000

678 **Wilhelm Stannat,** The theory of generalized Dirichlet forms and its applications in analysis and stochastics, 1999

677 **Volodymyr V. Lyubashenko,** Squared Hopf algebras, 1999

676 **S. Strelitz,** Asymptotics for solutions of linear differential equations having turning points with applications, 1999

675 **Michael B. Marcus and Jay Rosen,** Renormalized self-intersection local times and Wick power chaos processes, 1999

674 **R. Lawther and D. M. Testerman,** A_1 subgroups of exceptional algebraic groups, 1999

673 **John Lott,** Diffeomorphisms and noncommutative analytic torsion, 1999

672 **Yael Karshon,** Periodic Hamiltonian flows on four dimensional manifolds, 1999

671 **Andrzej Rosłanowski and Saharon Shelah,** Norms on possibilities I: Forcing with trees and creatures, 1999

670 **Steve Jackson,** A computation of δ_5^1, 1999

669 **Seán Keel and James McKernan,** Rational curves on quasi-projective surfaces, 1999

668 **E. N. Dancer and P. Poláčik,** Realization of vector fields and dynamics of spatially homogeneous parabolic equations, 1999

667 **Ethan Akin,** Simplicial dynamical systems, 1999

666 **Mark Hovey and Neil P. Strickland,** Morava K-theories and localisation, 1999

665 **George Lawrence Ashline,** The defect relation of meromorphic maps on parabolic manifolds, 1999

664 **Xia Chen,** Limit theorems for functionals of ergodic Markov chains with general state space, 1999

663 **Ola Bratteli and Palle E. T. Jorgensen,** Iterated function systems and permutation representation of the Cuntz algebra, 1999

662 **B. H. Bowditch,** Treelike structures arising from continua and convergence groups, 1999

661 **J. P. C. Greenlees,** Rational S^1-equivariant stable homotopy theory, 1999

660 **Dale E. Alspach,** Tensor products and independent sums of $\mathcal{L}_p$-spaces, $1 < p < \infty$, 1999

659 **R. D. Nussbaum and S. M. Verduyn Lunel,** Generalizations of the Perron-Frobenius theorem for nonlinear maps, 1999

658 **Hasna Riahi,** Study of the critical points at infinity arising from the failure of the Palais-Smale condition for n-body type problems, 1999

For a complete list of titles in this series, visit the
AMS Bookstore at **www.ams.org/bookstore/**.